The Human Face of Corporate Governance

To Lawrence
Thankyou for your
help and friendship

love,

Lynn

The Human Face of Corporate Governance

Lynn McGregor

palgrave

First published 2000 by
PALGRAVE
Houndmills, Basingstoke, Hampshire RG21 6XS and
175 Fifth Avenue, New York, N.Y. 10010
Companies and representatives throughout the world

PALGRAVE is the new global academic imprint of
St. Martin's Press LLC Scholarly and Reference Division and
Palgrave Publishers Ltd (formerly Macmillan Press Ltd).

ISBN 0–333–77205–9

This book is printed on paper suitable for recycling and made from fully managed and sustained forest sources.

A catalogue record for this book is available from the British Library.

Library of Congress Cataloging-in-Publication Data

McGregor, Lynn, 1944–
   The human face of corporate governance / by Lynn
   McGregor.
      p. cm.
   Includes bibliographical references and index.
   ISBN 0–333–77205–9
      1. Corporate governance. I. Title.
   HD2741 .M369 2000
   658.4'2—dc21   00–059179

10 9 8 7 6 5 4 3 2 1
09 08 07 06 05 04 03 02 01 00

Formatted by
The Ascenders Partnership, Basingstoke

Printed and bound in Great Britain by
Creative Print and Design (Wales),
Ebbw Vale

Contents

Dedications

This book is dedicated to all my current and past clients. I learnt everything I know about corporate governance from them.

It is also dedicated to my husband Lionel Snell. His professional help, patience, support and humor have been invaluable.

Acknowledgements

Special thanks to Pamela Ramsden with whom I have worked for many years.

Also to:
John Thomas for his research, hard work and suggestions.
Mike Woodhouse, for his experience, wisdom, inspiration and advice.
Jeff Westphall and the Westphall family.
Albie Sachs who told me to get off the phone and start writing.
Sir Christopher Hogg, Bob Monks and Bernard Taylor for their kind words.
Phoenix Graphics and Brian Platt for his cartoons.

Preface

This is an interesting and thoroughly worthwhile book. Its basic messages are relevant to any organization and any manager.

I was a classical scholar once upon a time and learnt early that the inscription above the gateway to the Delphic oracle, that touchstone of the ancient world, read 'know thyself'. To me, aged eighteen, it seemed banal advice. Now it seems to me the ultimate in managerial wisdom and I was glad to see that it heads the principles for self-governance in Lynn's fifth chapter.

Governance brings power; and power corrupts. The antidotes to that power are transparency, objectivity and accountability. All three require a steadfast and clear appreciation of oneself as seen by others.

The pay-offs for conducting governance in the way this book recommends can be enormous. This is more easily perceived by those who have jumped in and learnt to swim after their own fashion than by those who avoid the plunge. There are always admirable reasons for avoiding the plunge, of course. But this book is not one of them. On the contrary, Lynn's enthusiasm for her cause and her long experience of its benefits almost make one hear the trumpets sounding on the other side. I hope many will make the journey.

Sir Christopher Hogg

Sir Christopher Hogg is currently Chairman of the
Reuters Group, Allied Domecq and the National Theatre.

The human face is always changing. Sometimes it is happy. Sometimes it is sad. A caring expression can mask ruthless greed and ambition. A macho image can hide real love and concern.

What then is the human face of corporate governance? Or more important, what is it we desire to be?

Face also has another meaning. It is the side of a mountain to be climbed. It is the challenge we face to explore what it is to be truly human when governing corporations. As a result, this book focuses on the human side rather than on the financial and material aspects, although, of course, they are deeply connected.

Introduction

This book is written for those either involved in corporate governance or interested in its principles and practice, and who are concerned with the human side.

In my experience, most business leaders genuinely want to make the world a better place, and many do indeed achieve this in their own way. On the other hand, given how busy people are, it is seldom easy for them to stand back and reflect on the broader issues – even less to monitor the full consequences of their business decisions and see whether these are achieving all that was intended. Therefore, my intention in this book is to provoke thought as well as to suggest practical guidelines.

The book is intended firstly for existing or aspiring chairmen/women, chief executives (CEOs) and non-executive and executive directors. However, investors and other interested parties will also find it of use. As the book focuses on the faces of corporate governance, it roams freely between the technical aspects of governance and the more personal and social considerations gleaned from my work and discussions with board members and investors over the past 20 years.

The book is based partly on interviews with more than 200 board members about their views on corporate governance. It also takes into account many discussions that were held at working lunches with chairmen/women, CEOs and directors. Much of it stems directly from my own observations and perceptions when working with boards, and it is therefore highly personal in places.

Everybody has their own dreams, desires and concerns. Hopefully this book will give readers the opportunity to pause and reflect on their own views and experience, and to use some of the practical guidelines for their own benefit.

A personal perspective

As this book is specifically about the human face of corporate governance, the reader might appreciate an initial outline of my own views.

I feel very much a part of this earth. When I read a newspaper or turn on the television, there is news about what is happening in different parts of the world. On the one hand, when war is reported anywhere in the world, I can sense and almost smell the devastation. On the other, when a new cure for illness is announced, I feel moved to celebrate on behalf of those sufferers I know but also for every other sufferer across the world.

I spend hours e-mailing friends and colleagues across the world. In frustration, I struggle with new technologies, knowing that by the time I master one skill, another will supersede it. I feel pressured to do so many things that I have to fight hard to make time for friends and family. It is not always easy to stand back and reflect on whether my lifestyle is really serving me and those I love. I am also aware that I have a privileged and affluent life.

Indeed, I love the pace, but not the price.

When I think of the world, I think of faces – happy faces – unhappy faces. I see people leading lives of luxury, living in mansions, driving fast cars – some of them happy, others less so. I also see starving people struggling to make a living. In between are millions who live and work out their daily lives peace-fully and constructively.

Corporate governance touches us all. I buy gas from a filling station owned by a global company. My food is imported from distant countries and continents. When prices fluctuate or interest rates change, I am aware that those who govern our corporations (as well as members of government) affect the lives of millions, including myself.

When I think of my own role in this drama, I am aware of areas that are easy for me and that I do well. When I sit at a board meeting, I come alive if an issue being discussed is of interest to me. I become bored and frustrated if the conversation goes round and round with no conclusions. I get angry if directors provide inadequate information too late. In particular, I am stimulated and involved when options for the future are discussed or when a first-rate presentation is made.

I am also aware of the limits to my own knowledge. Above all, I am aware that there are many questions and very few easy answers. There are many paradoxes and conflicting interests that are not easy to resolve. Do I give money directly to the needy, or do I channel my surplus resources towards institutions that will increase general prosperity? Above all, I am aware that I am not perfect. I do not always get it right and have weaknesses that adversely affect others. For example, I am impatient, easily frustrated and can be blunt. This is difficult for some people. All I can do is constantly keep alert and continue to learn ways in which I can give of my best in spite of my own limitations.

And yet, when I watch others govern I am amazed by their knowledge, wisdom and expertise and their willingness to take responsibility for decisions concerning the fate of the corporations they govern. They are often addressing complex and fast-evolving problems as well as taking responsibility for organizations employing thousands of people, who in turn deliver products and services affecting millions more. In the case of global companies, whole nations and continents are being influenced.

I am deeply moved when I observe a chairman or a CEO who is not only able to generate profits, but who is able to make wise decisions that improve the lives of many. Conversely, I observe that many leaders are motivated mainly by status, greed and self-interest, and are purely interested in the 'bottom line' or getting their own way with no concern for the impact their actions have on others.

- If such people lead our corporations, what are the human consequences?
- How much does it matter?
- Is making money an end in itself, or a means to a better human end?
- If friends and family were personally affected by some of their actions, would directors take the same course?

Questions such as these lead me to believe that the human face of corporate governance is one of the most important issues for the new millennium.

This book will explain why.

SUMMARY OF CONTENTS

This book suggests that corporate governance is a highly complex art. The decisions made by corporate governors shape the destinies of corporations and influence the way in which the whole economic system is run. Corporate governance impacts on the quality of lives not only of shareholders, but employees and those communities impacted by key corporate decisions. This book stimulates thinking about the human aspects of corporate governance and offers practical suggestions for those who want to improve the quality of the ways in which they govern.

Part One explores what is governed and why the human side is important. Chapter 1 explores the social and technical climate that investors, board members and executive directors may have to govern in the not too distant future. Chapter 2 explains why it is important to take the human side more seriously. To do this can improve the ways in which leaders govern. However, it could well be the next stage in the evolution of corporate governance. If people can combine greater understanding and mastery of human processes with their business knowledge and expertise, they could significantly improve the quality of their own and others' lives. Improvement can take place at four levels – personal, interpersonal, intergroup and systemic. The book focuses on two levels, personal and interpersonal, as the foundation for the other two.

Part Two suggests ways in which personal governance can be improved. Chapter 3 demonstrates how different leadership styles influence the nature of corporate governance. Chapter 4 offers a simple working model of self-governance, followed by practical guidelines in Chapter 5.

The next section of the book, Part Three, explores interpersonal governance. Chapter 6 describes how different groups operate, while Chapter 7 offers a model for understanding how corporate decisions are made. Chapter 8 gives ideas for improving ways in which corporate groups work together. Chapter 9 discusses the shadow side of corporate governance and gives an approach and some suggestions as to how we can handle the destructive sides of our nature.

Finally, Part Four is a short section that links the book to the more formal roles and functions of main and supervisory boards and executive committees.

The summary and conclusions at the end of the book argue that improving the human side relates not only to those who are interested in the human face of corporate governance but also to all business leaders. It also refers to improving the quality of their working relationships. It ends with a suggestion that two issues are explored in more detail – the role of women in corporate governance and how to achieve better ways of being competitive and at the same time, more humane.

HOW TO READ THIS BOOK

Do not be daunted by this book. It can be read on different levels.

Reading the executive summaries will give a quick idea of what the book is about. If the reader wishes to understand the conceptual underpinning, this is usually found at the beginning of each part in the form of working models. The illustrations summarize the main concepts. If the reader is interested in practical guidelines, the best way is to skim through the book for those aspects of particular interest and to be highly selective. For the reader who is willing to be stretched and who would like a systemic way of improving the overall quality of the human side of their governance, going through the whole book can be highly beneficial.

Part One

The Meaning of Governance

1   What Is Being Governed?

Business prosperity cannot be commanded. People, team-
work, leadership, enterprise, experience and skills are what
really produce prosperity. There is no single formula to weld
these together, and it is dangerous to encourage the belief that
rules and regulations about structure will deliver success. (1)

▶ ▶ ▶   **EXECUTIVE SUMMARY**   ◀ ◀ ◀

In discussing the human face of corporate governance, the question arises as
to what is being governed. On the one hand this means how corporations are
governed and on the other it addresses the question of how key decisions
affect people. Predominant cultural values and beliefs directly influence us
all, and these may or may not be relevant to what we really want.
Technological innovation or environmental changes may also overturn the
cultural values and beliefs and our ways of living.

THE CHANGING SOCIAL CONTEXT OF CORPORATE GOVERNANCE

At present, we live in a society that values materialism, technology and
competition. On the one hand, these values encourage people to succeed, to
compete and take initiatives. They encourage vigorous commercial activity and
the success of the more enterprising companies. They also encourage innovation and constant progress. Often they ensure greater material wealth for more
people. However, as George Carlin, a well-known US comedian, put it, these
values do not always improve the quality of life.

These are the times of tall men, and short character; steep profits and shallow
relationships. These are the times of world peace, but domestic warfare; more
leisure, but less fun; more kinds of food, but less nutrition ... these are the
times of fancier houses but broken homes. It is a time when there is much in
the show window and nothing in the stockroom.

3

The last years of the twentieth century were about the beginning of e-commerce, cost cutting, redundancies, major mergers and acquisitions, and hostile takeovers. Developments in communications technology have enabled people to trade at greater speed across the globe, and this has created major opportunities.

Concerns have been expressed about the potential downside of the current system. On the financial level, people like Monks, Minnow and Soros (2) are now concerned about the dangerous consequences of irresponsible speculators who could undermine the economics of many countries and, indeed, the entire global economy. The United Nations Human Development Report (3) indicates that the gap between rich and poor, even in affluent countries, is widening. It has been forecast that wealthy countries will be able to invest in ever faster and cheaper communications. However, impoverished countries will not be able to keep up with constantly improving technologies and will be stuck with slower, more costly communications infrastructures. This will increase their isolation and poverty. There is also evidence of growing, rather than shrinking, inequality between the sexes, with still fewer women reaching the top of the ladder. The degree of this varies between countries, but the overall trend is there.

There is a great deal of concern about the potential effects of globalization. Trade liberalization might benefit global companies, but might also undermine the ability of governments to generate revenue or to remain autonomous. Rich companies are able to afford top lawyers and tax experts. They can also impose infrastructure demands on those countries in which they choose to operate. Not all governments can afford this without borrowing considerable sums with many strings attached. Thus, the end of the last century has seen the emergence of many anti-globalization groups with a whole range of concerns.

The concern extends to some of the practices of less developed countries. Some of these find it economically beneficial to allow activities that are unacceptable in the more developed countries. Child labor is a good example of this. So it is hardly surprising that in 1999 so many World Trade Organization (WTO) talks have collapsed. On top of the difficult issues relating to e-commerce and liberalization, WTO membership has increased to include many more poor nations. This diversity of needs, coupled with multiple vested interests, makes negotiation increasingly difficult.

All of this is part of an emerging world that everyone involved in corporate governance will have to grapple with. Changing markets and social patterns will also shape our ways of life and our assumptions about reality and how we treat each other.

Alexander Kobler (4), an eminent banker, identifies certain forces that he and others believe will change the world in the next 20 years. These include the following.

Expansion of the global market could limit the power of national governments

Technological advances will lower borders and reduce the cost of transport. 'In the future we will not only see companies and industries being optimized, but entire social systems as well … National states will lose some of their room for maneuver.' In the future, the role of governments will be both to provide environments in which companies can compete internationally as well as to 'minimize the social costs of globalization.' Regional blocks that 'will produce greater cohesion and a closer economic integration than the traditional nation state' will replace the power of individual governments. The challenge for countries will be to 'seek a path between independence and regional integration.'

Amount and speed of information

There will continue to be many innovations. Today's computers may evolve towards multifunctional machines so that a single box could serve as telephone, television and office service terminal. There is, of course, a counter-argument which predicts greater diversification – people might prefer small devices tailored specifically to their needs. Just as people will buy, say, a sports watch, a fun watch and a dress watch rather than one multi-purpose timepiece, so they may also prefer a television designed for the sitting room, a phone designed for wearing on the wrist and a computer designed for their study. All the same, they will need these devices to intercommunicate and operate in harmony. This means that the amount of information and the speed of communication will continue to increase. Thus, the market will indeed be dominated by new 'applications for digital networks, software and new media.'

Demand for intelligence

One banker I interviewed thought that in the next 20 years technology would really begin to deliver on its promises. Information in affluent societies will become cheaper than ever before, and the focus will then be on how intelligently information can be used. 'Knowledge and skill will in future be among the most important sources of comparative advantage in international competition.' There is likely to be so much information that 'a filter will be required which increases the ability to sift and evaluate it.' Intelligent information suppliers will be in increasing demand as 'many forward-looking jobs demand intelligence rather than detailed technical knowledge.'

New processes requiring less labor

In the future, Kobler predicts far-reaching advances in medicine, biotechnology and the production of new materials such as superconductors. Technological advances in digital modes of production mean that it would be possible to reduce the workforce involved in production to only 15 per cent of the total working population.

The size and nature of companies may change

'Direct communication channels make for a more active form of management.' There could be a move away from the present pattern of large mergers to smaller, more agile companies that can adapt more quickly to changes in the market and changing labor conditions. There will be greater division of labor and a trend towards more service and niche providers.

An increasing demand for environmentally safe products and services

Greater affluence will produce greater pollution. People will be looking for technologies that have a 'less harmful impact on the environment.'

A demographic time-bomb

By 2020, it is predicted that there will be a 50 per cent increase in the proportion of people over 65 years old, while by 2010 the 20–65-year-olds will have stabilized or even begun to decrease in affluent nations. Thus more people will be stopping work than starting. This has widespread implications for pension schemes and the funding of national social security systems. Kobler believes that innovative companies can help with this area. For example, healthcare, medical and security companies, and companies offering pension solutions will be much in demand. What is more, while the population of affluent nations is likely to dwindle, 70 per cent of the growth of the world's population is likely to be in the Third World. This in itself will change the ways in which the global population is viewed and what kinds of potential there may be in the marketplace.

A more mobile workforce conflicts with new processes requiring less labor

We have the technological capacity to radically diminish the workforce. In spite of this, because there will be fewer young people in the workplace, there is likely to be a greater demand for older people. As a result of this demographic time-bomb, 'working models will need to be adjusted for both women and older people.' Increased telecommuting work could 'trigger massive adjustments in corporate structures and processes.'

If any of these predictions come true, there will be serious implications for investors, main and supervisory boards and executives. Investors will be drawn to new market sectors and different regions, with corresponding effects on the gap between rich and poor. Non-executive directors will be faced with unfamiliar technologies and different ways to access information in unfamiliar formats. Financial services will continue to evolve, and executive directors may be faced with a shifting workforce with changing demands in their workplace and lifestyle. Automatic retirement at a certain age, for example, may become a thing of the past as we find ways to enable people of all ages to develop and move with the times. As the population grows older, will it grow wiser?

Only time will tell how many of these predictions come true. What is certain is that the world of industry will not be the same in 2020 as it is in 2000. There will be changes, and times of change are times of both opportunity and danger. In particular, it presents an opportunity to create a better world. Alexander Kobler, however, points out that although there will always be natural disasters and political instability, 'the most unpredictable factor continues to be people who shape the future with their ideas. All the more reason for businesses to participate actively in the process as early as possible.' He points out that it is the 'strategies embarked upon today that will decide whether companies succeed or fail in the time period we are considering.'

It is, therefore, not just the commercial and financial factors in corporate governance that need to be considered. Until the human side is seriously addressed, we will be deluding ourselves about our ability to manage, adapt to and control our environment. We simply cannot be certain what effect these changes might have on us and how we will react as human beings. We can only learn to ask the right questions.

- If the best solution is one that will make the world a better place financially, socially, emotionally and spiritually for the greatest number of people, can decisions made purely in financial terms really be the way to achieve this?
- At what stage does self-interest impact negatively on the lives of others?
- Are the people who lead and administer global organizations qualified to make effective decisions as wise human beings?
- How can one achieve a balance between creating and encouraging change and monitoring destructive behavior?
- If people do not basically care about each other, how does that affect the quality of governance?

These are the sorts of questions to be raised in times of change. They set the scene for the next chapter, which presents the case for the human face of corporate governance.

Notes
1. Hampel (1998) *Hampel Report on Corporate Governance*, Gee Publishing.
2. Soros, G. (1998) *The Crisis of Global Capitalism*, Little Brown and Company.
   Monks, Robert A.G. (1998) *The Emperor's Nightingale*, Capstone Publishing Ltd.
   Monks, Robert A.G., Minnow, N. (1996) *Watching the Watchers*, Blackwell Publishers Inc.
3. *The Human Development Report* (1999) Oxford University Press for UNDP.
4. Kobler, A. (1999) *2020 – The challenges facing the next generation*, UBS Swiss Outlook.

2 The Human Face of Corporate Governance

It is too easy to forget that corporate governance is for human beings and is carried out by human beings.

▶ ▶ ▶   **EXECUTIVE SUMMARY**   ◀ ◀ ◀

This chapter sets out the conceptual framework for the book and focuses on the fact that governance is carried out by human beings.

It begins with a definition of corporate governance and is followed by reasons as to why the human side is so important, before addressing the question: who actually governs? To answer this question, corporate governance is analyzed at four distinct levels – systemic, intergroup, interpersonal and personal or 'self-governance'. To define the overall context, all four levels are discussed briefly in this chapter, while the rest of the book focuses mainly on self and interpersonal governance.

The function of governance is to rule, lead, create and maintain structures and systems and to monitor performance. How people govern depends upon their values and beliefs, their ability to make decisions, as well as their capacity to ensure effective implementation of decisions. Decisions that directors take affect the destiny and quality of the lives of many human beings so that, although expertise, knowledge and skills are essential for efficient governance, something more is needed.

Who a person really is, or what groups of people really are, will determine the quality of decisions and the impact of those decisions on other people. The key question is not so much whether people are following a particular code of practice. It is about whether those who govern consider themselves responsible for and accountable for the impact that their decisions have on other people. Many directors are deeply concerned that the choices they make benefit as many people as possible. Others are more occupied with their own self-interest.

As creative human beings, we have the capacity to be both constructive and destructive in the ways in which we shape our own lives and affect the lives of others. This is just as relevant to those who govern corporations as it is to those who govern nations.

WHAT DOES CORPORATE GOVERNANCE MEAN?

Many people believe that corporate governance has been reduced to mere rules, regulations and codes of practice, and that it fails to address the real purpose which, according to the Hampel Report (1), is to create wealth.

Corporate governance is also associated with the work of main or supervisory boards that are supposed to represent the interests of shareholders and, in some countries, stakeholders as well. They are also expected to monitor the performance of executive boards or committees, and to make ultimate decisions concerning the fate of companies.

Although these are all aspects of governance, I would suggest that they are functions of governance, and not the essence. Governance of corporations is not dissimilar to any other form of governance, but because the word means different things to different people, the following working definition is suggested for this book:

Governance is the process whereby people in power make decisions that create, destroy or maintain social systems, structures and processes.

Corporate governance is therefore the process whereby people in power direct, monitor and lead corporations, and thereby either create, modify or destroy the structures and systems under which they operate. Corporate governors are both potential agents for change and also guardians of existing ways of working. As such, they are therefore a significant part of the fabric of our society. This is often forgotten in the business of making money and responding to a competitive market.

WHY IS THE HUMAN SIDE IMPORTANT?

Those in power have the potential to improve or destroy the quality of the lives of many people.

As creative human beings, we all have the potential for constructive or destructive behavior. All leaders have the potential to use or abuse power. It depends on the conscious and subconscious choices we make.

More so than ever, however, business leaders have the power to affect the quality of life of large populations. Their decisions directly affect the physical environment, the goods and services we use, and the way we communicate with each other. Their belief systems pervade our working and family lives. If a family is dependent on two parents working, and the work environment is not sympathetic to families, then the children will suffer. The ways in which those involved in corporate governance value themselves and treat other human beings really matter. It is no longer enough for a business leader, be they a board member or a shareholder, to say that their sole purpose is to make money or to generate profits. Who they are and what kind of people they are impacts on the lives of many.

In working with business leaders, I have witnessed many examples of inspired governance. I have experienced interactions on boards where the quality of trust and working relationships have been deeply fulfilling, and where the effects have been outstanding at every level – financial, social and emotional. Those people who have helped create, with a group of like-minded people, an environment in which outstanding performance takes place, have all said that they would never forget how wonderful it was. 'We know each other well. We deal with real issues and come up with workable and highly satisfying solutions.'

The potential is there for us to live enriched and fulfilling lives as whole human beings. In the last century we began to master the art of global communication. In the affluent world we have the techniques and knowledge to produce products and services and to run financially effective organizations. According to Peter Russel (2) we also have the capacity to end starvation across the globe.

We equally have the capacity to make our own and others' lives diminished and miserable. This is often the result of unconsciousness, short-term action and the failure to recognize our own and others' shortcomings in time. It is also the result of lack of foresight about the implications of our decisions or actions on others.

Many corporate leaders who are adept at managing the material aspects of governance prove to be underdeveloped in other areas, such as, for example, the spiritual and emotional sides. Those who are interested in the so-called 'soft' side of life are often not as good at generating material wealth or as interested in the financial aspects of life. However, there are enough examples to indicate that the marriage of 'soft' and 'hard' can produce a quality of life that is at once materially, spiritually and emotionally superior to that produced by one side on its own.

We have a greater understanding of human nature than ever before. This understanding has mainly been used in crude ways by the business world. Lists of competencies and categorizing people in boxes help directors to understand the standards expected of them at any one time. Listing people by category is useful in understanding individual differences and indicating whether there is balance in a group. However, there is more to human nature than these approaches can reveal. When people govern, they apply their talents and perceptions, ways of making decisions, wisdom, understanding and creativity. They also live either consciously or unconsciously according to their values and beliefs about themselves and their perception of reality.

Human nature is highly complex and cannot be categorized in boxes. However, a greater understanding of human nature, and ways to foster and harness the best in people, can only enhance the quality of governance. As most people involved in governance have already proved their professional expertise in material terms, there is often resistance to spending time on the softer side of human nature. Some leaders feel such concerns are self-indulgent. However, knowledge and expertise about the less tangible aspects of human nature could be added to existing business skills without undermining their capacities to produce goods and services and to generate wealth.

Our definition of wealth could easily be expanded to include human elements. For example, how far do business decisions take into account deep-seated human emotional, spiritual, intellectual and physical needs? Manfred Max-Neef portrays a deep understanding of universal human needs which relate to those involved with governance as much as to any other human being. They are also relevant for those governors who are concerned that they meet as many human needs as possible in the course of their work. They will be summarized here and discussed in more detail later in the book.

Is it the duty of corporate governors to take into account basic human needs?

Manfred Max-Neef (3) outlines nine basic human needs: subsistence, protection, affection, understanding, participation, identity, idleness, creation and freedom. He suggests four practical ways in which these can be described meaningfully: having, being, doing and interacting. These can be measured against basic needs. In this way, it is possible to check whether certain needs are more satisfied than others, and which demand attention.

There are different ways in which needs can be satisfied. He suggests that 'synergic satisfiers' are best because they meet more than one basic need at the same time. If significant satisfaction does not take place, it can damage both individuals and groups of people. If a business leader is suffering from a basic unfulfilled need, his or her behavior could well be driven by neuroses that undermine the people with whom he or she directly works. Such a person could also neglect the needs of the wider populations he or she serves. This will be discussed in more detail later in the book.

The corporate governor who says that he or she is not involved or interested in the human side is not only making a statement about how little they value people as human beings. They are also making a statement about themselves and the face they present to others. Financially successful people who measure themselves only in terms of status and money have an impoverished and limited idea of who they truly are and what they could really create.

If those involved in corporate governance were to take into account the less tangible aspects of human nature more consciously in their work, it could make a significant difference to the quality of their own and others' lives. Throughout the book, examples will be given to indicate the importance of the more human face of corporate governance.

WHO GOVERNS?

> *The most difficult thing is to tackle the institutionalized*
> *abuse of power. There is no single person to deal with.*

It is all very well to make a plea for integrating the less tangible aspects into corporate governance, but whom are we addressing? Who actually governs?

The conventional idea of corporate governance is that it takes place at board level. Many books have been written about the role of the chairman and the non-executive and executive directors. This is one aspect of governance, but not the whole story. Corporate governance is a complex system that takes place at many levels.

There is a conspiracy theory to the effect that a very few extremely powerful people control global business. In the sense that captains of industry do talk with each other, sit on each other's boards and share major decisions, this may hold some truth. However, in another sense it could be argued that everyone working in the corporate field is involved in governance. Each one of us has a direct or indirect role to play in creating or maintaining the corporate system. Every time a key decision is made that makes a difference, a form of governance is taking place. In many cases, those who formulate and implement decisions, *de facto*, are different from those with formal, legal, *de jure* responsibilities. However directly or indirectly anyone is involved in governance, they are likely to operate within specific areas and levels (Figure 2.1).

Fig. 2.1   Levels of governance

Governance acts on several levels. The most pervading level – described below as the 'systemic level' – operates through numerous institutions, many not directly related to each other.

Social changes may well begin as random fluctuations, but as people respond to these changes, either they become institutionalized into the system or else the system itself is changed – as in the case of e-commerce. Systems are often maintained or changed through power struggles between different institutions. How much consideration is or is not specifically given to human implications depends on the key players' values and levels of consciousness.

Action groups often bring specific issues – such as environmental threats – into public awareness. These in turn influence the decisions that leaders make. There is often an assumption that a trade-off is needed between commercial and human interests. But does this need to be the case?

Because of the complexity of the ways in which the system operates, it is difficult to generalize about who specifically operates these broader systems. It is easier to be more specific about those groups who are clearly seen as key players in the field of corporate governance. So the next level – 'intergroup governance' – will be discussed, followed by 'interpersonal' and, finally, personal or 'self-governance.'

Systemic governance

'Systemic governance' relates to the overall system within which corporate governance takes place. It has to do with rules, laws and regulations concerning acceptable behavior, trading agreements, tax etc. It recognizes the social, economic and technical environment in which we operate, at both a global and national level. Usually it operates through recognized institutions such as governments, stock exchanges, and the World Trade Organization (WTO).

Recently, a number of people have questioned the effectiveness of the free trade system which, they believe, is dangerously volatile, and have suggested that a global monitoring institution be set up to ensure a sustainable system. (5)

This level is about how the overall business system operates, and defines the legal and fiduciary responsibilities of directors. People who operate at this

---

SYSTEMIC GOVERNANCE

The type of pension scheme offered by an insurance company might seem a purely financial concern – it is, after all, the customer's choice – and yet it can have serious repercussions at the systemic level.

During the boom years of the late 1990s, money purchase pension schemes were heavily promoted. These are programs where the final pension is based on the value of the fund invested in at the time of retirement, as opposed to a final salary scheme where the pension is based on the salary at time of retirement.

Too great a reliance on money purchase pensions raises a serious problem. If the stock market plunges, then a whole generation of retirees could be left with severely depleted income for the rest of their lives. This in turn has serious repercussions on public spending. (4)

level are often involved in the formulation of new rules and regulations. Some take a watchdog role and monitor or analyze movements and trends that point major investors towards new markets.

At this level the thinking is often purely statistical. An example of this was a fatal rail crash whose cause was put down to low safety margins. A spokesman for one of the companies involved stated: 'Well, for any disaster like this one you have to trade off lives with cost.' At a lunch discussion soon after, a number of the company's directors clearly felt that if customers were offered the choice of a more expensive transport system involving safer signaling systems, they would choose the less expensive option.

There are no easy answers to such dilemmas. However, it does raise the question of how do we, as a society, value the cost of human lives? Current values and beliefs affect what people are prepared to do or not to do. These beliefs and values can be changed in many ways, for example by such actual or potential disasters as global warming, or again by radical advances in technology, the influence of epoch-making writers or philosophers, or even a small dedicated band of highly motivated people who create tangible successes which are then copied.

The point is that corporate governance is not only about the business of making money for shareholders. It is also about the quality of life we can live as human beings.

Intergroup governance

This concerns the relationships between specific groups that help to maintain or create national, international or global systems, regulations and structures. At the global or macro level, these include regulatory bodies such as governments, the Competition Commission, the WTO etc. They also include pressure groups such as trade associations, trade unions or environmental groups who can apply pressure to change policies. In this arena, the media plays a powerful part as they influence thinking about what is going on.

The typical working relationship between groups is that of formal representatives negotiating on behalf of their institutions. Outcomes are often determined by how successful people are at making deals, and how good they are at playing power games. These power games can directly affect the way in which the system operates. In the United States, for example, one such struggle has led to business leaders dominating local government. The decision of many US states to permit people trading through e-commerce not to pay tax will directly erode the revenue of those states. A longer-term solution will require significant changes to the US tax system, and it is far from clear whether a win-win solution favoring both sides will be devised.

In any one corporation, the groups usually associated with corporate governance are shareholders, including major investors, main or supervisory boards and executive committees/boards. Investors not only affect the amount of money invested or disinvested, but also tend to influence company timescales. Some put pressure on companies to deliver short-term results at the expense of long-term sustainability. Many directors complain that short-term investors undermine their company's long-term planning. 'They are so interested in quick bucks that they destroy the potential of what could be produced in the long term.'

There is a difference of opinion as to whether shareholders or major investors should be directly involved with corporate governance. Some react only on financial performance. Others, such as Lens, a successful investment company, intervene directly and apply pressure if they feel crucial changes need to be made.

Main or supervisory boards are supposed to act in the interests of shareholders and to ensure that there are good returns on investment. In some countries such as Germany and Japan, they also represent the interests of other stakeholders such as banks, trade unions and employees. Main boards have the ultimate authority to make major decisions on behalf of shareholders (which often include themselves) concerning the fate of companies. In many companies they have both a monitoring and an advisory role in terms of decisions made by the executive board or committee. They also directly influence the selection or resignation of the chairman, CEO and other directors.

Because boards stand between the executive group and the shareholders, they have to steer a careful path. Although non-executive directors are meant to be objective, this is not always easy. There can be many conflicts of interest; for example, if a director holds shares in more than one competing company or a subsidiary of a company that may be sold off, it will affect the way that director makes decisions. Directors may also feel loyal to a chairman or CEO when it might not be in the interests of the company for that person to remain. The relationships between the main board and the executive board or committee can therefore be fairly complex. Some boards merely rubber-stamp decisions. On other boards, the chairman will encourage full debate of issues and will expect non-executive directors to add value to executive thinking. Yet others are more interested in the value of buying and selling, acquiring or merging companies. The orientation of any board will fundamentally affect the way a company is governed. Crucial to board performance is the chairman, who sets the tone and style of board dynamics. The relationship between the chairman and CEO is also vital.

Executive boards or committees, led by the CEO, are expected to formulate strategic proposals and, once strategy is agreed, to produce expected results –

particularly financial targets and with no nasty surprises. These are the results that shareholders depend upon for good returns on their investments. Both the main board and investors need to have confidence in the ability of the executives to deliver.

Relationships between these three groups are crucial for the system to operate effectively. In cases where they are working well, people are usually quite happy and get on with the business of generating wealth. However, there are also many examples where lack of understanding, trust and respect, combined with inadequate information and hidden agendas, erode the possibility of these groups working well together.

A typical example is when a major investor feels that a chairman and/or CEO is not telling them the truth. Sometimes it is a feeling backed up by some circumstantial evidence. A CEO presents a rosier picture than the figures show. The investor becomes suspicious, and tends to challenge the chairman and

---

INTERGROUP GOVERNANCE

Mergers and acquisitions raise a clear example of intergroup governance. What may seem a perfect marriage in the eyes of the financiers and lawyers brought in at the start of a merger, can turn out a nightmare to management and human resources staff left to clear up the mess.

The problem is often a question of 'company culture' – a tenuous concept much trumpeted in the marketing and recruitment literature, but easily overlooked in the context of hard business.

The clash of cultures is perhaps to be expected when the merger crosses continental borders. At the time of writing, Ford and Mazda appear to be struggling to see eye to eye, with James Miller quitting the post of president after only two years – the second president to resign abruptly from Mazda, despite record profit expectations in 1999. But even without such obvious national differences, the Glaxo Wellcome and Smithkline Beecham merger collapsed on account of what was described as 'management philosophy, corporate culture and differences of approach.'

Intergroup governance issues need not be solely between companies. When BP Amoco wanted to merge with Atlantic Richfield there was opposition from the US Federal Trade Commission (FTC). The FTC feared that the merger of the two largest oil suppliers in the West Coast region could lead to higher prices. So the companies worked directly with the governors of Alaska and California – two of the key states affected – to reach a satisfactory agreement. According to BP Amoco CEO Sir John Brown, agreement with Alaska was a 'critical step' in securing US regulatory consent. (6)

CEO, sometimes aggressively. The CEO and chairman become defensive, and a vicious circle is created.

It takes only a few adverse media reports on the company for investment to be withdrawn. In some cases this is justified, but not in all cases.

Interpersonal governance

In my experience, actual decisions are not often made solely in the formal arena. Instead, people tend to gravitate towards like-minded colleagues who have similar vested interests. Decisions are often negotiated off-line and rubber-stamped at formal meetings. How much the human implications are discussed during these private meetings is difficult to determine.

Decisions may be formally ratified at an institutional level, but are actually defined by the ways in which individuals relate to each other. How each person contributes, their thinking, experience and expertise influence the quality and timeliness of decision-making. The quality of governance depends, therefore, on the individual and combined input of participants.

When I ask board members about the best experiences they have had as board members, their faces light up. Most describe the experience as stimulating, enjoyable and where 'the combined decisions are better than each person's input.' 'The quality of debate is excellent.' 'We know each other well and get on well together. This means we can thrash out issues without pretending or hiding anything. We get through more this way.' 'The atmosphere is lively, awake and there is often great humor.' 'There is a spirit of working together for a common purpose rather than fighting each other.'

Good working relationships are not only necessary for effective governance, they are also satisfying and fulfilling for those engaged in the task. Although good relationships do not in themselves create effective governance, those groups who have the professional expertise, intelligence and competence to fulfil their tasks, and also have excellent relationships, do produce outstanding financial performance. They often also have a reputation for treating people with respect, and they attract and maintain good people. A chairman of two major global companies was delighted with the annual results. 'They are all due to improved communication.'

In one highly successful software pioneering company, people were encouraged to take initiatives and responsibility for influencing decisions. Many people felt frustrated because they felt as if a lot of hard work was being done with numerous committees, but nothing was being achieved. Some people left to go to other companies. Yet over two-thirds of those who left asked to come back. 'It is a difficult company to work for, but I feel more valued as a person here. In spite of the frustrations, it is also more interesting than being stuck in a bureaucracy.'

The human side is often expressed in value or mission statements. They tend to summarize what people feel is important. 'We want to be number one worldwide and work in an open and honest way in a friendly and sound environment in which all people are fully empowered.' Although 80 per cent of mission statements are similar, people attach importance to them because they express shared values. Setting up mission statements is one thing. Living them, however, is another.

Dysfunctional groups seldom have value statements and, if they do, very seldom apply them. There are numerous examples of ineffective governance. Dominant chairmen or CEOs stifle contributions and initiative. Directors competing inappropriately with each other take up discussion time with backbiting, internal politics and putting each other down. People complain about each other in private. There is lack of trust and tension. Angry directors sabotage meetings by not allowing things to go forward. People go round and round in circles because they do not want to commit to a decision. Functional tasks are not handled well.

INTERPERSONAL GOVERNANCE

Investor confidence in the UK oil and gas exploration company Lasmo was 'dented' when managing director Chris Wright suddenly left – according to the financial press.

Problems had been brewing for some time. Exploration companies had been demoted in the FTSE listings from having their own sector to being part of the Oil & Gas category. To investors they now looked like small fish beside giants like BP Amoco. Lasmo had also built a reputation as a competent seller of assets but a poor buyer, after their acquisition of Monument Oil and Gas.

These problems unfortunately built up after the arrival in 1998 of Chris Wright from BP, who set about installing a new organizational structure on what had been reputedly a very easy-going business. CEO Joe Darby was known for his more relaxed management style in contrast to the allegedly restless and impatient manner of Mr Wright, which led to 'love it or hate it' divisions amongst staff and colleagues.

This is the sort of 'oil and water' mix that can create a perfect balance when handled correctly. But when other problems intervene and there simply is not the time and energy to devote to interpersonal governance, then the 'marriage' falls apart. Chris Wright left, and Joe Darby, according to some analysts, could have less than six months to save the image of a company which had been a star performer in the FTSE 100 just three years before. (7)

In most mediocre groups, the human side is seldom discussed and few clear policies agreed. The board is often fragmented, low in morale and shows signs of dissatisfaction.

Personal governance

Whether a group of people work well together or not is largely dependent upon the attitude and caliber of individuals in that group. We are often so busy being busy that we forget that the ways in which we govern our own actions, thoughts and feelings directly affect the ways in which we govern. Who and what we are as individuals, as well as collectively, is a key foundation that takes the human face of corporate governance into account. Two leaders, identical in their technical ability to govern a company, will produce quite different results according to their individual natures.

For example, there were two companies in the same industry, both highly successfully financially, with both chairmen highly creative and entre-preneurial. The first was run by a dictator who used fear as his power base. His motivation was to be in power at whatever cost. He believed the company could not survive without him. He used to phone people up at five o'clock in the morning and tell them how useless they were. His highly paid senior managers were afraid to challenge him. They spent time and energy hiding from him. The CEO was extremely unhappy and kept talking about resigning. He felt he had to stay out of loyalty to his people as well as safeguarding his pension. His wife was clinically depressed. One son was in trouble. There was a history of nervous breakdowns in the company. When the chairman retired there was no one to replace him. The company was leaderless for a number of years and lost its edge.

The other company chairman demanded equally high standards, but encouraged and acknowledged good work. Through his enthusiasm, he inspired his directors to do more than they believed they could. He had a constructive relationship with his CEO and executive group. He ensured that the non-executive directors took part in stimulating debate about key issues. He was fascinated by learning about new things and how to improve the ways in which he worked. People queued to join his company. When he retired, quite a few people were ready to succeed him and the transition went smoothly.

In reality, people are complex and have both strengths and weaknesses as well as different attitudes, values and beliefs. Everyone is unique, with a different background and style. They bring their own ways of thinking, experience and expertise. Everyone has both specific talents and blind spots. They also have different needs and ways of relating to others. Those who are not fully aware of their strengths and talents act out of instinct and may not be exploiting what they have to the fullest. Powerful people who are not aware of their weaknesses, or how they might be abusing their power, are less likely to find ways of compensating for their shortcomings.

Many board members do not really believe in self-development, as they think that to spend time on themselves is selfish. However, a common trait in leaders who have a reputation for being great leaders is that they have a high level of self-awareness. They are also prepared to admit to and to work on their weaknesses. One chairman of a global company summed it up as: 'Nobody is perfect. At most, we are about 60 per cent. It is my job to help our executives deal with the missing 40 per cent. I myself need feedback too from time to time to check that I am on the right path.'

Most people in positions of power are not perfect. In fact, I have never met such a paragon. The higher up people go, the greater the talent and, often, the greater the faults. The ways in which people govern themselves have an effect on the quality of their own and others' lives. This even extends to what goods and services each of us chooses to buy. Our own values and attitudes about others and ourselves have a profound effect on the human side of what we as individuals do. Personal governance is therefore a major component of corporate governance and will be discussed in greater detail in the next section.

SUMMARY AND CONCLUSIONS

In this chapter it has been suggested that corporate governance is a complex interrelated system. The system is influenced by dominant cultural values and beliefs about business and society. It is also affected by technological and social changes. Certain trends were described as potentially affecting corporate life.

It was argued that corporate governance is highly developed in terms of the financial and technological aspects of business. The next stage is to include more consciously the humane, less tangible aspects of governance. How this is actually done depends on the levels at which people involved in corporate governance operate. Control through main or supervisory boards is only one aspect of corporate governance. Corporate governance takes place at four different levels: the level of the overall system (systemic), the level at which governing groups formally relate with each other (intergroup), the inter-personal level and, finally, personal governance.

This chapter has given the context in which governance takes place. The next part of the book looks at some of the less tangible components of corporate governance and suggests ways in which individuals and groups can be more effective at different levels. The levels that apply to everyone are personal and interpersonal governance. Because of this they will provide most of the material in this book.

Notes

1. Hampel (1998) *Hampel Report on Corporate Governance*, Gee Publishing.
2. Russel, P. (1995) *Global Brain Awakens*, Element Books.
3. Max-Neef, M.A. (1991) *Human Scale Development*, Apex Press.
4. *Financial Times* (19 October 1999) Pensions and the stock market slide.
5. Soros, G. (1998) *The Crisis of Global Capitalism*, Little Brown and Company.
6. *Sunday Business* (16 January 2000) Don't forget the human side of mergers and acquisitions.
    Financial Times (16 December 1999) Ford and Mazda bump wheels.
    Financial Times (3 December 1999) BP Amoco finalizes Arco deal details.
7. *Financial Times* (16 December 1999) Investors find Lasmo no oil painting after boardroom dust-up.

Part Two

Personal Governance

3   Corporate Governors are Human Beings: Three Examples

Know thyself.
Know thine enemy
Know the ground and the terrain
And thy victory will be assured. (1)

▶ ▶ ▶   **EXECUTIVE SUMMARY**   ◀ ◀ ◀

In Part One, a wider context for corporate governance was given with an argument for the importance of the human side. Four levels of governance were described, with two common to everyone involved in governance – interpersonal and personal. Part Two of the book focuses on the importance of personal governance. It is argued that the particular style, strengths and weaknesses that any individual brings will determine the nature and quality of the ways in which they govern.

To illustrate the significance of this, examples will be given of three very different people. As will be explained, they are not real individuals, but composite characters drawn from my experience.

The first is the story of a successful woman banker, the second that of the Norwegian CEO of a printing company who is also chairman of a distribution company. The third tells of a Chinese man who runs a global IT company. All are different personalities with different values, outlooks and talents. This is reflected in the ways in which they tackle their roles as significant business leaders.

What motivates individual corporate governors?
How does a better understanding of self help to improve the general effectiveness of governance?

In the past, most highly effective business leaders did excellent work by getting on with the job without consciously thinking about what they were doing and how they were working.

In the latter part of the last century, with the development of psychotherapy and the human growth movement, more people have recognized the value of greater self-awareness. Corporate governance is effected through collections of individuals. What an individual decides about a business not only affects people's material lives. It also influences the ways in which people are treated.

The more aware you are of how you operate and what motivates you, the greater the possibility that you can recognize and build on your strengths as well as finding ways of modifying your weaknesses and blind spots. No one is perfect. Everyone has feet of clay. The challenge is to recognize and to deal with those less attractive aspects of oneself so that they do not get in the way. On the other hand, everyone has unique areas of genius. The challenge, also, is to find ways of harnessing these to their best advantage.

This chapter gives a thumbnail sketch of three different kinds of leader. Chapter 4 will suggest a model for looking at aspects of personal governance so that readers can start thinking about how their styles affect the ways in which they govern. Chapter 5, Guidelines for Personal Governance, concludes the section on personal governance.

The following stories are invented. There are two reasons for this. The first is that all the profiles of business leaders that have been made over the years contain personal and detailed life histories and information that is highly confidential. The second reason is that during the course of profiling over four hundred people certain patterns have emerged. The subjects therefore, while being unique, also have some characteristics that many business leaders have in common. No doubt the reader will be able to identify with some aspects of a character or recognize certain traits in people they know.

PROFILE OF JANE HUNTER

CURRENT ROLES

Chairman of Bank International Plc.
Non-executive director of three hi-tech companies.
Trustee of a major cancer charity.
Age 47. Married. Mother of three.

PERSONALITY

Sociable. Curious about people. Enjoys working with small intimate groups. Sense of fun.
Highly energetic, optimistic, extremely bright, fascinated by how things work.

From an early age, Jane has always been interested in where things come from and where they go to. Has an intuitive ability to understand trends and to visualize the future.

Able to listen to views of others and accurately summarize what has been said.

NEEDS

To feel she can trust people, and to work in a caring, harmonious environment.

Does not like surprises.

VALUES AND BELIEFS

Business is a competitive game. The purpose is to win and to be ahead of the competition.

Expedient investment and talented management will always get you ahead.

Honesty, dependency and trust are prerequisites of good business.

KNOWLEDGE AND EXPERTISE

History major, Oxbridge. Captain of hockey team at college. Gained a first at Harvard as an MBA student. Qualified analyst. Successful investor. Joined Bank International as a senior manager in the early 1980s, became CEO at the age of 38, and was elected chairman at 45 (first woman and youngest chairman).

Helped turn the bank from a national to an international player.

Expertise in marketing and strategy. Technologically knowledgeable, particularly in the field of IT.

DECISION–ACTION

Ability to understand the implications of trends and to envision innovative outcomes. Ability to scan the environment and to create windows of opportunity, to see possibilities and to move extremely fast. Will set commercial priorities and persuade people to take risks. She can see the whole picture from a multi-dimensional perspective as well as being able to operate practically on many levels at the same time.

Good at checking out intuition with analysis. Does her homework well.

LEADERSHIP STYLE

Inspires people to see what is possible, to take risks and to win through. Insists on high intellectual and conceptual standards and encourages high-level debate. Will listen with respect to all opinions, sum them up, and then lead a group into arriving at a creative decision. Extremely sensitive to mood and morale, she provides a stimulating and safe environment for trust. She attracts people of exceptionally high caliber.

MAIN STRENGTHS

Reputation for integrity. Long-term strategic thinker. Vision of how large organizations could achieve competitive advantage through flexibility.

Understanding and improving the value of companies.

Successful investor, not just in companies, but in talent also.

Very good when things are going well.

MAIN WEAKNESSES

Does not like to criticize people, and will avoid confrontation.

Administration. Following things up. Patience. Suffers fools gladly. Over-trusting and naïve.

Jane Hunter: A vignette

I first met Jane when she had been chairman of the bank for two successful years. Three of her non-executives were due to retire and she was busy selecting new ones. 'I believe that chemistry is vital. We're looking for mature and extremely bright people. Not necessarily technical, but aware and open to dealing with fast change and understanding the implications. I want stars, but not prima donnas. I want people who can work together in a collegiate atmosphere. That's the sort of chemistry I want.'

Once her board was in place, it certainly lived up to her aims. Board members all said that they looked forward to attending meetings, that they enjoyed the dialogue and reaching 'decisions together that were greater than the sum of the parts.' She was clear that 'the critical element of a board is that there is a degree of trust, allowing free and frank exchange. It's worth spending time away to get to know each other. No one should be afraid that what they say could damage their position – they should welcome having their opinions challenged. But if anyone tries to control and manipulate directors, they will get their come-uppance. I can't stand ego games.'

The bank began a period of dramatic growth and significantly increased market share. It looked set to become the world leader in on-line and automated banking. Jane was able to use her unique talent to understand the implications of technology, business and social trends as part of a historical continuum. While others reacted to change, she could read the signs, anticipate the consequences and had a track record for getting ahead of the competition. She was one of the first to recognize on-line financial services as both a threat and an opportunity to traditional banks. But she knew that, to succeed, a rapid change in thinking and culture in the world of banking would be necessary. Part of this was the ability to enable large companies to move fast and to keep up to date with technology.

It looked like a job well done as share prices rose steadily. Investors were pleased. But barely 18 months later, the whole enterprise was in the doldrums, with rumors in financial centers of a pending financial crisis for the bank.

What went wrong?

A year into the new board, it became evident that her CEO was failing to meet financial targets and was falling well short of budget. He and his financial director sprang a nasty surprise at one board meeting that shocked her. 'I do not know how I could have been so deceived,' she said.

The problem was that she had appointed a CEO who was a relatively unknown outsider. Appealing to her innate élitism, the man had a brilliant academic record but little business experience. When it became evident he was underperforming, she failed to intervene – even though the executives had repeatedly asked for her help. She became concerned that the bank's financial problems were threatening her ambitious IT strategy. Her bank was beginning to lose its lead.

The non-executive directors were unhappy, though they still had plenty of good things to say about the chairman. 'She demands high performance and clear thinking, and keeps people on their mettle.' 'She is good at picking holes in arguments in a constructive way, that motivates people rather than puts them down.' 'She is stimulating and good fun to work with'. 'I respect her enormously, because she speaks her mind' … and so on.

On the other hand, 'She gets very upset when there are tensions and conflicts.' 'She just refused to intervene when the CEO had clearly gone astray, when the situation could have been saved much earlier on.'

Her view was: 'I strongly believe in not intervening in the CEO's executive responsibilities. I would intervene only if the company was seriously at risk. Otherwise, I expect him to deliver. My role is to offer a wider, more objective overview. I am there if he needs to bounce ideas off me.'

Fine words, strongly expressed. But they concealed the real problem: In discussing the situation, it become increasingly clear that Jane Hunter was simply afraid of confrontation. Her success was bred upon success: with her sheer competence, intelligence and positivity she was a dynamo for generating success. In this positive context she would be challenging, competitive, a real business warrior. Under such an influence, things seldom went wrong, but when they did a different side of her emerged. The chairman who sparred so successfully with her own brilliant board could not face taking her CEO to task for non-performance.

'My father was an ambassador who traveled a lot. I didn't like the lack of continuity, but he taught me the importance of respect and clear judgment. My mother was gentle, good fun and had an astute understanding of people. I inherited from both of them.' However, 'the problem was my father used to get drunk and lash out at my mother. He could be very cruel and violent at times. I was terrified of him and learnt to hide when he started drinking.' At an early age she made a decision that 'things must never be allowed to "go wrong" like that. It simply cannot happen.' She put a great deal of energy into building a harmonious environment around her. When things were going well, she could be really tough – in fact she could be hard on people without realizing it. But when things went badly with people she felt close to, she tended to ignore the problem.

She was highly aware of her shortcoming but did not know what to do about it. She realized that she had to find a way to confront her CEO in a positive way without bringing up old memories and patterns of behavior. It took her considerable courage and persistence to overcome her fear and to develop skills to deal with situations that demanded confrontation. However, she did learn to work with the CEO in such a way that she could determine what went wrong without causing him to be defensive. He had not been good at monitoring the performance of his managing directors and had let things slip without knowing it. She was then able to suggest to him ways in which he could improve and develop. He followed her advice and the company rapidly showed signs of recovery. With great relief, she and her board were able to go back to what they did best – helping the executive to create wealth in such a way that they are always at least two steps ahead of their competitors.

PROFILE OF LARS CARLSEN

CURRENT ROLES

CEO Global Print Services.

Chairman EuroTransit. President of the Association for Promoting IT Knowledge in Europe and Scandinavian countries.

Married with three grown-up children. Age 57.

PERSONALITY

Perfectionist. Highly determined. Fascinated by simplifying things and seeing basic patterns in complex situations. Tendency to go for one thing at a time.

Extremely enthusiastic, but can lose his temper easily when frustrated. Action-oriented and motivated to complete tasks once started.

NEEDS

'To prove to myself that, together with my people, I can overcome any opposition.'

Honesty and fairness from others.

Difficult challenges.

Constant stimulation and variety.

VALUES AND BELIEFS

You have to be tough in a rat race or you will go under.

Perseverance is extremely important.

It is better to be honest and unpopular than deceive people.

Fairness.

DECISION–ACTION

Extremely precise and detailed knowledge of a situation. Able to analyze data at speed, especially financial information, and to pinpoint inconsistencies. Simplifies and arranges information in terms of priorities and order of importance. Quickly compares options and builds a good case for the best decision. Commits to action and then directs activities in order to get things done.

KNOWLEDGE AND EXPERTISE

Engineering. Ability to understand how things are constructed, how they could be done, and what is needed to keep them going.
Operational experience.
Astute sense of the market. Ability to inspire highly competitive sales forces to gain market share.
Starting, building and maintaining businesses.
Turning round businesses at speed.
Spearheading and championing ideas or projects.
Presentational skills. Popular with the media.

LEADERSHIP STYLE

Strong leader. He listens to views, crystallizes them, and then comes up with an idea that turns into a decision. After that, 'let's get on with it and make it happen.' He also expects his directors to do their homework. He will analyze the data, interpret it, 'get to the nub of the problem' and pinpoint what could stop them from winning. He is also excellent at providing a strong argument that is well thought out and inspires people. He exudes confidence and the certainty that even the most difficult challenges can be overcome. People either love or hate him. One of his directors said: 'If you submit to him, then you are finished. If you work side by side with him, then we are both unstoppable.' He is also good with large audiences. He gives simple, clear messages and motivates people to do their best.

MAIN STRENGTHS

He draws a clear line between his roles as CEO and chairman. This means that there is no ambivalence or subtle power games.
He will not tolerate destructive 'politics.'
Ability to simplify confused and complicated situations or ideas.
Extremely good at getting things done and producing good results.
He is prepared to be open-minded and fair if he is challenged.
Insists on and rewards high standards.
Tough and successful negotiator.

MAIN WEAKNESSES
Speaks before he thinks, and sometimes upsets people.
Impatient. Can lose his temper easily.
Without knowing it, he can dominate or intimidate others easily.
Insensitive as to how other people may be feeling.
Wants to move fast without realizing that different people move at different speeds.
Gives the impression that he is always right.
Over self-critical about his failings.

Lars Carlsen: A vignette

I first met Lars when he took over an ailing conglomerate that had been leaderless for two years. It took him 18 months to turn it round. 'It was so diverse, that the front did not know where the end was.' He had a deserved reputation for rationalizing, streamlining companies and insisting that they 'stuck to their knitting. No company can excel at more than two things at the same time.' His policy was to use consultants to find out where the company had key strengths, and then exploit them. 'We sold off the rest or made people redundant – in a fair way, of course.' He then put together what he called an A-team. He had himself and his directors assessed so that he could fully utilize their strengths as early as possible. 'Together we mounted a campaign to meet fairly challenging targets.' This also meant insisting on higher standards across the board. 'It was not a company for the faint-hearted. We did a roadshow to let people know exactly what was expected, and the rewards they would get if they performed. We also told them straight out, "if you can't stand the pace, then this company is not for you."' His goal was to put the company in order, and to create an organization in which people 'walked tall.'

He had had a tough childhood. Brought up on a small farm, he had a taste of greater things as Norwegian schools ice hockey champion until stricken by polio at age 16. He still walks with a slight limp. Denied further sporting glory, he turned to engineering – financing his studies by harvesting seaweed for the cosmetics industry. He sold the business for a considerable amount. 'People complain and call me negative and critical. But engineers are trained to home in on flaws. That is a major contribution that I make. People don't understand this.' He made his mark as a civil engineer leading major international projects – one small African nation owes a substantial part of its commercial infrastructure to a road-building project he managed. He was instrumental in growing his company 500 per cent during the 1970s.

I next met him in his capacity as CEO of a major international print corporation. He was also chairman of a pan-European distribution group, and non-executive director of a packaging business and a barcode data capture company that supplied goods and services to his major interests.

It was the mid-1990s. He could remember the times when the printing industry was full of hot metal and heavy machinery. As a perfectionist with an aggressive sales approach, he pushed the business to new heights against fierce competition. What's more, he was a brilliant administrator and multi-task juggler who could successfully lead his company into an increasingly digitized age. However, as one director commented, 'he never let us forget that we are simply here to produce and sell paper with ink on it.'

He commanded considerable respect. As one director put it: 'I know exactly where I stand with him. What you see is what you get.' He was recognized as a brilliant fixer of crises and an intuitive entrepreneur. No one doubted his loyalty, nor his basic humanity – he had won prizes for his role in establishing recycling centers that generated profits for local charities.

So it was very difficult for the chairman to have to tell Lars that he had to change his style. 'Too many people are complaining that they don't like working with you.' He had started in an industry with strong traditions where he had been among like-minded people – tough, aggressive and competitive street fighters. He produced stunning results, so his macho manner had been accepted in most cases as the natural right of the 'alpha male.' But now he was in an industry with a substantial information technology bias – staffed by people from a different culture with specialist knowledge. It does not go down well to storm unannounced into the office of a Ph.D. in computer science and demand that they tidy their desk!

After the initial shock to his ego, and a game of golf, Lars decided that the chairman was right. However, he also felt that although he was prepared to do what he could to change, he was unlikely to change his upfront, blunt personality. The solution was fairly obvious. He hired a general manager who could handle the day-to-day running of his business, which left him free to concentrate on the outward-looking areas where his competitive nature was much needed.

He found it difficult at first to let go until outside interests diverted some of his energy. He became deeply involved in the expansion of his distribution business, and began to make his mark as a media spokesperson for increasing IT awareness in the European business community.

Back with his executive directors, he found the right balance. 'They get on with their own thing and ask me to point out the flaws well in advance. They don't seem to mind it any more if I am blunt or bring urgency to the business. In fact they like it. We now have a good laugh about it.' Happily, with the late 1990s economic upturn, the financial results proved so good that even Lars had to admit that the right mix of cultures 'can be commercial dynamite.'

PROFILE OF WANG LEE

CURRENT ROLES

President of W.L. Plc.

Honorary Fellow and Professor of three universities.

Chairman of a UN Peace and Regeneration committee for East Timor.

Married with four children and two mistresses.

Age 47.

PERSONALITY

Quiet, extremely observant, constantly amused by life and not easily thrown off course. Self-deprecating and difficult to get to know. He is good at establishing contacts and putting people in touch with each other. He is always ready to help friends or contacts if they are in need. Combines astute opportunism with compassion for the underprivileged. He has sponsored over two hundred young people and been instrumental in their careers without them knowing. He is also a significant but anonymous donor to Amnesty International.

NEEDS

Must have information about what is going on in the market. Needs to be in touch with what is happening in his companies.

To be constantly working with a variety of people.

To be in at the heart of any crucial negotiations.

VALUES AND BELIEFS

It is important to be circumspect, and watchful as well as diligent. If you know what is going on, you can make the right choices.

It is better to curb desire and passion if you wish to understand the patterns of life.

Invest in people, but do not take possession of them.

KNOWLEDGE AND EXPERTISE

Brilliant at analyzing financial results, seeing anomalies and giving accurate guidance to companies.

Understanding of how to reconfigure major alliances and acquisitions into growth and profit centers.

Natural understanding of the applications of new technologies, and an ability to optimize current opportunities worldwide.

Able to encourage innovation and commercial ideas, and to enable enterprise to thrive across a number of companies. He puts groups of people in touch with each other so that together they can be more formidable.

Able to assess whether new ventures are working out, and, if they are not, to withdraw before any damage is done.

Good at supporting profit accountable companies and at the same time providing corporate values and brands that people are proud to identify with.

DECISION–ACTION

Good at conceptually understanding where business opportunities can be exploited. Lateral, out-of-the-box thinker.

Highly developed sense of timing, and when it is time to move he makes and maintains important contacts and sounds out ideas with potential customers as well as with his directors before coming to conclusions.

Delegates to others to make things happen.

Monitors in an unobtrusive way.

LEADERSHIP STYLE

Known as the 'invisible' leader. He tends to have regional boards and to intervene only when needed or when he wishes to introduce innovation.

He allows plenty of scope and room to thrive, and people repay this many times over.

He is well known for his ability to listen and for his compassion.

MAIN STRENGTHS

Brilliant opportunist and negotiator.

Good at networking, making and maintaining relationships.

Shrewd and wise.

MAIN WEAKNESSES

Unpredictable.

Difficult to get to know.

Not good at seeing when his key people are struggling.

Wang Lee: A vignette

People often asked me if Lee actually had his own office. Meetings with him were invariably at others' premises – boardrooms, restaurants, airports or, most typically, in a comfortable, beautiful and relatively unknown hotel lounge.

If he wasn't speaking excitedly on a mobile phone, he was hard to spot – a small, modestly dressed man who usually identified his contacts first and watched with an amused twinkle in his eye as they gazed around in search of someone who looked like the leader of a gigantic and reputedly voracious communications empire.

If pressed about his modest and mercurial style of business, he was likely to claim that 'the wise chairman is one who acts without action, and accomplishes much without taking credit.' He attributed much of his success to his Taoist background, and claimed that it stood him in good stead in a life that constantly took him round the globe.

His restless lifestyle reflected his early displacement. An only child born in China at the end of the Second World War, his family fled from the Communist revolution and finally settled in Singapore. His father found employment as a waiter – a far cry from his former glory as a head teacher, but quite acceptable to a man of great modesty and firm Buddhist principles. It was a different story for Lee's mother, an ambitious woman who preferred the teachings of Confucius. She never really recovered from the loss of family ties and dignity, and placed all her hope in her son. 'I take after my father who is more self-effacing and has always maintained his dignity. My mother gave me ambition to succeed. She has a very strong sense of family ties, and we have a very extended family that spreads across the world. Networking is a natural way of life for me.'

He had a reputation for networking and successful deal making. As a young man in Singapore, he did well at school, and then took a degree in accountancy, but became bored with it. 'It gave me a good understanding of figures, but I wanted to make money, not just count it.' So he joined an international consultancy firm as a financial consultant and soon had an international network of influential friends. He also established lasting contacts in the Singapore government. 'I know everyone whom one needs to know in this part of the world.' He attributes his success to establishing and maintaining key relationships. 'It takes time, but it is well worth it.' He was headhunted as a CEO for a smaller rival, and within three years it had outstripped his old firm.

In the seventies friends laughed at him because he did not think much of computers in those days. And yet he was to be proved right in the long term because that was the day of the stand-alone mainframe and he would arrogantly argue that 'until those things learn to communicate like people they are nothing more than a high-speed abacus.' He was an immediate convert to the fax and mobile phone, however, and when the first personal computer network was announced he became fascinated by the possibilities of electronic communications. The company he founded soon had network nodes in every continent. 'We slashed the cost of phone calls and faxes, and pioneered the way for global data networking.'

Part of his expansion was to move into North America in the late 1990s. He reached outside his familiar network and recruited Silicon Valley 'whiz kid' Leroy Steenman as CEO to consolidate his US/Canadian operations. On the face of it, it was another smart move because Lee had picked up the first tremors of the impending Asian crisis. He sensed the importance of the North American continent for future growth, and the need for an all-American figurehead while he quietly got on with his international deals.

It was by no means easy at first. Leroy knew Lee by reputation, and was in awe of his achievements, particularly as the company rode the Asian crisis in the 1990s. But once they started working together, he found Lee's approach to business difficult to deal with. 'He won't let me intervene, even though I tell him that some of our companies are not doing well. I have no idea what he is thinking. I find Lee devious and shifty. I never know where I stand with him.' Leroy seriously considered leaving because he was so frustrated. What he did not do was to share his unease directly with Lee.

I met Lee when he had an intuitive feeling that not all was well. He felt that a third party might help too. Once he realized it was a cultural problem, Lee thought 'the best way someone can understand me is to see me in action.' He invited Leroy to accompany him on one of his extended trips around the world. It took Leroy some time to understand the effectiveness of Lee's style of invisible leadership. Lee explained his philosophy. 'Ruling a large corporation is like cooking small fish – the less you stir it the better.'

Lee achieved considerable success because he had avoided the worst of the Asian crisis through foresight, sound financial planning and by building sound regional bases surrounded by autonomous companies with a strong global identity. 'My people are profit accountable. But we all know each other well. This means that when there is a cross-border opportunity, small groups of people get together to work creatively to create new business opportunities.'

He believed strongly in the cross-fertilization of ideas. When he saw the Asian crisis coming, he immersed himself in traditional Western business practice and absorbed many of its techniques in his natural eclectic manner: 'A large corporation should be like the low country towards which all streams flow.'

Leroy found the trip 'highly educational and fascinating. But I will never be able to operate like him.' It took him a few weeks to realize that 'Lee is totally inscrutable, and does not give much away; he always keeps his word. People out there trust him.' What Lee was not good at was sorting out cultural differences and tensions between people. He was too busy scanning the environment for the next set of opportunities.

What Leroy did was to set up facilitated workshop sessions in the different regions and invite the CEOs and MDs of different companies to discuss what they were doing and to air any difficulties they had.

It was a lot of hard work. People at first complained that it was a 'waste of time' and they were 'too busy right now.' But at the time of writing, most of them would now agree it has been worthwhile. The group as a whole is dealing better with cultural problems, while the US operation has achieved some remarkable successes through applying more subtle opportunism against its competitors.

Most significant of all, the US company has become a major player in establishing a firm business base for US companies in China by leap-frogging the competition. It has won contracts, in spite of tough competition, because, as one Chinese official stated, 'There is understanding of our needs and successful adaptation to our culture. It is a pleasure to work with them.'

SUMMARY AND CONCLUSIONS

Three completely different kinds of people have been described in this chapter. The purpose has been to illustrate that the ways in which governance takes place depend a great deal on who and what each individual is. It is just as important as their technical and financial expertise. These stories, along with other descriptions, will be used to illustrate certain aspects of governance in the book. The next chapter distills what I have learned about individuals over the years. It offers a model for self-governance that the reader can use to gain greater insights into themselves, their colleagues or their competitors.

Note
1. Griffith, S. B. (1971) Sun Tzu (5th century) *The Art of War*, Oxford University Press.

4   A Model for Self-Governance

▶ ▶ ▶   **EXECUTIVE SUMMARY**   ◀ ◀ ◀

In the last chapter three stories were used to illustrate how individual personality and style can have a profound effect on the way one governs and on one's contribution to corporate governance. This is self-evident yet easy to forget when one is caught up in day-to-day business pressures. So this chapter takes a closer look at the individual personality, its components and scope for change.

The same kinds of question emerge again and again when working with individual chairmen/women, CEOs and non-executive or executive directors.

How do I know that I am really doing the right things?
How could I improve what I do?
How does my work impact on others?
Am I getting the best out of people?
Am I doing all I can to make the greatest positive difference?

It is difficult for any one person to answer these questions on their own. The best starting point is to gain greater self-awareness, and then to use that awareness to improve the quality of what one does.

Self-knowledge is an act of continuous learning and an important part of the dance that each of us has with life. We are continuously challenged and invited to test our assumptions, whether they are about ourselves, about others or life in general. The conclusions we draw, whether correct or not, inform the actions we take and this in turn alters the reality around us.

George Soros referred to this process as 'reflexivity.' He suggested that 'Regimes have two aspects: the way people think and the way things really are. The two aspects interact in a reflexive fashion. The mode of thinking influences the actual state of affairs, and vice versa.' (1)

In the same way that it is not always easy to distinguish between reality, creativity and illusion, it is equally difficult to understand the complexities of human nature, especially one's own. Despite degrees in psychology, sociology and philosophy, literature and drama and over twenty years of work experience, I must confess that I do not know all the answers as to what really motivates a person.

There is an argument that 'every baby is born with circuits that compute information enabling it to function in the physical world. As this brain develops, there are systems in our left hemisphere that automatically enable each person to interpret the world and to construct belief systems. Our automatic brains accomplish the amazing feat of constructing our sense of self and our past – a process clearly fraught with errors of perceptions, memory and judgment.' (2) It is this 'sense of self' however flawed, that is the first step towards improving the quality of governance.

One of the ways I have found by which people develop greater understanding is through the use of models that provide a systematic, if simplified, analysis of complex concepts. The model I propose for self-governance (Figure 4.1) is based on profiling and observing over four hundred board members. Although I know that it works, I make no grandiose claims for it. It is merely hypothetical, provisional and intended to provide a frame of reference for the reader to use to gain greater insights about themselves and others.

What you do stems from who you are. Your actions are the external manifestation of your intrinsic qualities. As they are interconnected, it is difficult to separate one from the other. For example, decision-making includes thoughts, values, beliefs and feelings. It also often ends in action and the achievement of specific results. However, for the purpose of greater clarity of understanding, the model is divided into two areas: 'who you are' and 'what you do.' Each area is further subdivided into facets. I have chosen to illustrate it in the form of a diamond because it has a core nature, which is 'forever.' We are, however, shaped and polished by our decisions and life experiences. We become multi-faceted like a gemstone.

'Who you are' outlines some of the intrinsic qualities within a person. It begins with the core nature – those characteristics that are hard-wired and unlikely to change. How your personality develops is, to some extent, dependent on what you learn from your life experiences. This in turn influences the values and beliefs by which you live. All of these affect the ways in which you make decisions and choices. Decision-making is a combination of using what you were born with together with what you learn in the course of your life. These four components, your core nature, life experience, your values and beliefs and the unique ways in which you make decisions, direct your actions and influence the effect you have on others.

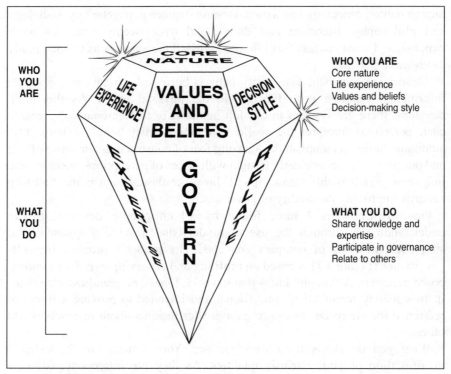

Fig. 4.1   Model for self-governance

WHO YOU ARE

Core nature – enduring traits

> *Each mortal thing does one thing and the same:*
> *Deals out that being indoors each one dwells;*
> *Selves – goes itself; myself it speaks and spells,*
> *Crying What I do is me; for that I came. (3)*

Core nature refers to a person's inherent characteristics. If you look at a photograph of yourself at one year old and compare it with one of yourself today, you now look older and different, but the same person is there. Core nature is related to a person's DNA, that is, their basic inherited characteristics. It is what is hard-wired in us, what is ascribable to 'nature' as opposed to 'nurture.'

The traditional analogy is to consider an acorn. Its core nature is to grow into an oak tree, but whether the tree turns out to be part of a forest, or a single splendid landmark, or even a gnarled specimen in a bonsai collection – that is another matter. In the Nordic tradition, Ygdrassil is the symbolic Tree of Life, continuously attacked by weather and eaten by animals and insects. Whatever is done to it, it remains the same tree, even though from time to time people may lose touch with its central truth.

The three characters in Chapter 3, had they been real, would have had different core natures – though you might have to know them very intimately to be sure how much was innate and how much had been learned. Jane was naturally bubbly and bright. She also had a natural historical perspective that was further developed at university. Lars was highly action-oriented and competitive, and had a natural need to complete things. Wang was introverted and reflective by nature. He had an instinct for seeing how things connected, and a natural fascination for how things work in practice.

Core nature is neither positive nor negative – unless one is genetically psychotic. It has the potential to be both constructive and destructive. The good or bad that people do is more a consequence of how well they engage in the dance of life and what they make of the basic clay from which they are formed.

In listening to the life stories of over four hundred board members and observing them at work, I find that certain behaviors appear constant, regardless of the situations and circumstances. They form the basis of patterns of behavior that a person repeats over and over again and so provide a clue to the person's core nature. This does not deny that people also grow and change through learning.

These elements of the core nature include the following.

- **Basic personality traits** that are inherent in a person. For example, although many of us can learn or be inspired to be altruistic, some leaders are genetically hard-wired to be altruistic. (4) It also includes such things as mannerisms, whether someone is naturally extrovert or introvert, a natural disposition to be practical or intellectual, and the nature of one's smile.

- **Predisposition to a decision-making style.** Each person seems to have a natural inclination to make decisions in certain ways. For example, some people, like Jane, tend to jump in without looking. Others are naturally more cautious and need to examine everything in detail. Innate differences in decision-making appear to be connected with patterns of energy that are deeply entrenched in the personality. (5)

- **Innate intelligence.** This does not refer to IQ-type intelligence. There are some well-defined skills that are naturally and instinctively easy for a person, and others that they lack however hard they try. For example, some people can spot what is wrong about a proposition or activity instinctively, while others spend hours looking for the gaps and never find them.

- **Inherent talents.** This is akin to an innate intelligence, but far broader in scope. If a person is asked to recall four peak events of their lives, times when they have felt they added real value, then I have discovered that there will be common ingredients in each event – a particular pattern of excellence that runs through a person's life when given the opportunity to manifest itself. This talent can remain instinctive or it can be developed, and it may be appropriate at certain times but not others. For example, Winston Churchill had a natural predisposition to lead the country to victory in time of war. He was less good at running a country in time of peace.

- **Natural blind spots and weaknesses.** This is the converse or negative equivalent of the previous positive elements. Although certain weaknesses and blind spots come from lack of experience, or learned behavior, others do seem to be innate – for example, Lars' predisposition to lose his temper, or the above-mentioned inability to see what is wrong with a proposition.

These aspects of personality seem to be hard-wired and can only be significantly changed by chemical or neurological influences, including extreme trauma. These are the foundations upon which character is formed as we build upon what we are born with and respond to life's experiences.

Life experience

This section is about the influence of 'nurture,' while the previous paragraphs addressed our 'nature.'

Experience is the raw material for learning how to adjust to life and participate in the acts of creation or destruction. Experience can be positive or negative, happy or painful. It includes every type of learning, be it active or passive, practical or conceptual or emotional.

In listening to people's life histories, it becomes apparent just how much significant life experiences affect their basic beliefs. This is especially true when the experience takes place in early childhood, but we can be as deeply affected by highly significant events later in life such as a first love affair or a brush with death.

The purpose here is not to insist that, in selecting a director, we must look beyond the CV and examine every incident in his or her life. It is simply to remind us that underlying a plain CV lies a wealth of experience that is at least as valuable as any technical qualification. Also, from the individual's point of view, there will be times of blockage or crisis when it may be necessary to gain a better understanding of their own history and its effects in order to make progress. This can be a chance to recognize and adjust unsatisfactory patterns of behavior.

If someone needs to re-evaluate themselves, this understanding of what they have learned from their background, and how it agrees or conflicts with their core nature, can be a useful starting point.

Personal life experience

In the formative years, the most common influences quoted were family, siblings, school, particular teachers, holidays, illnesses and accidents. Some people were also affected by particular circumstances, for example whether they were poor or rich, whether they were adopted or their parents divorced and whether the country in which they were brought up was at peace or at war.

In the examples from Chapter 3, Jane was affected by having an alcoholic father. Lars was deeply influenced by the farming community he was brought up in. Wang believed he owed his interpersonal abilities to get on with people to being a refugee.

I myself attribute many of my own positive and negative characteristics to a childhood in South Africa where my parents were anti-apartheid activists (6). From them I learned a deep enjoyment of life and an ability to question the status quo. I also acquired high levels of anxiety as well as the ability to scan the environment for actual or potential danger. Thus, my ability to help chairmen and CEOs to avoid potential risk has been the positive outcome of a difficult experience.

At a later age formal education also plays a part. School, university, college or business schools may have a fundamental influence on one's character. One non-executive director attributed his ability to think systematically to a 'very strict attention to order and routine when I was at boarding school.'

Indeed, very few of the people I work with have ever stopped learning, for an eagerness to learn is a key quality of the most successful people that I have met. Life, work, marriage, family relationships and friendships have, as one director put it, 'enabled me to grow as a person.'

Another important aspect of learning has been mentioned by everyone that I have interviewed. It is the value of mentors at formative periods of a person's life.

Without exception, everyone that I have profiled has identified key people who have been crucial in their careers – people who have taken a positive interest in the person, mentored them and helped them in the advancement of their education or career. There seems to be a pattern where mutual recognition and respect enables someone to go further on their path than they would have on their own.

The converse can also hold. For some people the biggest source of learning has been a tyrant or 'tormentor.' someone who made such a negative impression that they reacted or took a stand against what they considered to be unacceptable behavior. This will be considered later when we look at the 'shadow side of governance.'

Life experience also includes the accumulation of knowledge and expertise gleaned not only from general life experience, but also through education, training and work experience. Any board member brings a wealth of specialist knowledge, understanding of finance, people and organizations. Hopefully, they also bring qualities such as wisdom, ability to take risks and integrity.

Values and beliefs

> *Thou oughtest therefore to have put my money to the exchangers, and then at my coming I should have received mine own with usury.*
>
> *For unto everyone that hath shall be given, and he shall have abundance; but from him that hath not shall be taken away even that which he hath. (7)*
>
> *Without interest, there would be no money to lend. (8)*

Without ethical standards, people have the excuse to do what they like to other people. Those business leaders who command the most respect consistently live by strongly held values and beliefs. For example, integrity is one of the values most sought after in the business world. Indeed, it is argued that without trust, it is difficult to conduct business that is mutually beneficial. Other values often mentioned are compassion, wisdom and responsibility. Another quality often not mentioned, but obviously evident in people who have made significant contributions to the greater good, is, as one chairman put it, 'the capacity to give and receive love.' This word is often banned in business circles, but often practiced. Where it is not practiced, it might do well to remember the words in the Bible.

'I may be able to speak the language of men and even of angels, but if I have no love, my speech is no more than a noisy gong or a clanging bell. I may have the gift of inspired preaching; I may have all knowledge and understand all secrets; I may have all the faith needed to move mountains but if I have no love, I am nothing.' I would argue that the same applies to the generation of great wealth at the expense of other people – *'But if I have no love, this does me no good.'* (9)

Values and beliefs are a combination of natural tendencies and learning. Our principles and beliefs provide a framework within which to decide the difference between right and wrong, particularly relevant in terms of the human face of corporate governance.

We absorb our values and beliefs from parents, teachers, childhood heroes and the culture we live in. They may be instilled, or reinforced, by the life experience discussed in the last section, but they may be quite independent or even contradictory to that experience. For example, a father who owes his success to extreme arrogance might drill his child to believe that humility is the greatest virtue.

Whatever their origin, these beliefs shape our response to life and our positions concerning corporate governance. Everyone has personal values and beliefs, some of them conscious and explicit, others unconscious or even denied.

Jane Hunter, for example, believed firmly in the value of honesty and trust – she had a well-deserved reputation for integrity. On the other hand, she had an unconscious belief that she would be in danger if she confronted someone. Wang, on the other hand, believed in leading through allowing others freedom to act and not pushing his own views. He assumed that others were like him and could not believe it when his American CEO was uncomfortable with this. Lars Carlsen believed that it was important to complete projects, and so he invested a great deal of energy and steamrolled others to get results.

These deep-seated beliefs may be formalized as ethical codes – the rules and laws that draw a line between acceptable and non-acceptable behavior and provide a frame of reference for the meaning we give to life. All major religions and philosophies provide such guidelines for living. However, even people who do not subscribe to any such formal set of morals will have their own conscious or unconscious framework.

In the course of my work, I have not come across a single person who did not have deep beliefs and values. Often these were hidden until the opportunity to explore and articulate them arose. Everyone has their own interpretation of reality. Deeply religious people typically believe in a higher being that provides guidance as to how they should live, while atheists often want to pack everything they can into their lives – 'there are no dress rehearsals.' I was deeply touched by a holocaust victim who said that he did not know what to believe: 'It does not make any sense, what happened. All I can do is see to my garden, which means focusing on doing the best I can for my family and in my work.'

Our moral and ethical values are like a compass that steers a person through life. Indeed, the sense of 'meaning' behind decisions is a vitally important yet often overlooked driving force. When what one does in life is no longer meaningful, in this sense, then life can feel purposeless and impoverished, however rich one may be. See the example of the advertising agency MD who suffered 'the pain of loss of meaning.'

THE PAIN OF LOSS OF MEANING

A chairman of a powerful advertising agency complained that his managing director was 'losing it.' 'He has lost his sparkle.' He was coming into work consistently late and staring vacantly out the window during meetings. 'It's as though he has lost interest in his work.' People were beginning to notice it, and were going to his superiors to discuss important issues instead of to him.

When asked what was going on, the man complained of boredom. 'I've been doing the same thing day after day for over ten years. This no longer has any meaning for me. The work is trivial. It doesn't improve anyone's life – apart from the dollars. I spend most of my waking time producing transient activities that have no enduring value. What am I doing with my life?'

This typical mid-career crisis offered him a golden opportunity to re-evaluate his values and beliefs. But when he thought about changing his job and his life, he could not detach himself from the perceived belief that he needed to make more money. He already had enough for his family, future and lifestyle, yet he could not bear being less well off than his peer group.

Unfortunately for him, he stayed and continued to be dispirited until he was asked to take early retirement. A colleague who visited him a few years later was shocked. 'He wanders around his house in slippers spending all his time watching chat shows and sport on the television.'

Not only are these values and beliefs the stars by which we navigate through life, they also guide us through the human side of corporate governance. Is the face we present compassionate and humane, or does it seem as if we only care for self-interest?

In Chapter 2, I defined governance as the creation, maintenance or destruction of organizations. Corporate governance is more like an art form than a science. One definition of art is 'the exploration and expression of meaning through a medium.' (10) The medium of corporate governance is the world of business. If this is so, a question for anyone responsible for corporate governance is, 'Through your contributions, what are you trying to create, maintain or destroy, particularly in terms of your role as a human being?'

THE INSPIRATION OF CLEAR VISION AND VALUES

The young president of a high-tech company was highly idealistic. 'I want to make a real and positive difference to the lives of my family and the people with whom I work. I also want to make a difference worldwide.' He worked with both the board and the executive committee to co-create visions 'that we really want to achieve.'

It took a lot of hard work and time for both groups to agree on a few core values that everyone felt that they could commit to and put into practice in their work. Key values included 'mutual respect, innovation, creativity and self and team development.' They also took on 'to make a positive difference for the benefit of the wider world.'

Three years later, when asked about the vision, values and mission that had been formulated, every person lit up and was able, with total enthusiasm, to share them and the benefits of living up to them.

One director described the experience as follows. 'It was not easy at first. We struggled with what these words meant and how to live them. I had to let go of telling people what to do. In the short term, it took longer than I wanted. But now my people are working so well on their own that my job is almost redundant. I have developed enormously as a person, and feel I have the inner resources to deal with almost anything. What we are doing is enormously worthwhile, and we have already helped to change international trading agreements to be fairer to smaller companies. Our products are people-friendly and are more popular than most of our competitors. Our business partners tell us that they want to work with us because they feel respected, and enjoy the spirit in which we do things as well as the quality of the work we produce.'

The atmosphere in the company is positive and purposeful. Newcomers are able to describe the predominant values of the company without needing to read or be told anything. The firm has a high reputation for integrity. It is also extremely successful financially.

The important point is the way that an influential leader's values and beliefs can rub off on the people they govern. A dictatorial chairman is likely to dominate and subdue directors while believing it his or her duty to provide firm leadership and decisiveness. Conversely, a CEO who believes in involving many people in decision-making is likely to encourage greater democracy in the workplace.

Some values remain constant. However, there may be times when one needs to re-examine them. A life crisis or major opportunity can change a person's beliefs about life or other people.

One chairman I knew used to be essentially trusting of people until he discovered two of his directors engaged in insider dealing. It came as a bombshell; he felt he had to offer his resignation and it took a long time before he could trust anyone again. He had to give up his belief system that 'all people were basically trustworthy' and learn to discriminate.

Another chairman had quite the opposite experience. He was brought up to believe that most people were crooks, and had to revise his opinions when two of his colleagues supported him through a difficult time.

Values and beliefs about people will also influence the way an individual treats a larger population. Unless people involved in governance are prepared to take social responsibility for what they do, their role as corporate governors is questionable. I remember asking an investor whether it mattered to him that disinvesting from one country would cause terrible poverty. 'That is not my problem. My job is to do what will pay the shareholders the best interest.' Fortunately, however, I have found that business leaders who are genuinely concerned about the well-being of those that they govern form the majority.

As Demb and Naubauer (11) pointed out, one of the greatest challenges for people who govern is how to handle paradoxes. For example, if it is clearly more profitable to sell or relocate a company, despite devastating implications for the staff and community, how can the human face be reconciled with the financial aspect? Rationalizations such as 'in the long run more people will benefit from a more profitable organization' do not really address the human side because they treat people as populations rather than as human beings. If the distinction is not clear, imagine trying to put that argument to a tearful mother of four who has lost her livelihood. As one director commented, 'Taking the human side into account makes decision-making even more complex and difficult.' This may be so, but as decision-making is one of the key tasks of corporate governors, can it be ignored?

As values and beliefs play such a vital part in governance it would be a good thing if every person involved in governance could take time to reflect on their values and beliefs: whether to reaffirm values they still hold dear, to question beliefs that may no longer be valid, or to reflect on the extent to which they have been living out their values and beliefs. The next chapter will provide some guidelines for those who would like to take stock of this aspect of their lives. This will also include suggestions for dealing with conflicting values and interests.

Values and beliefs play a large part in the next section, decision-making. Who you are also reflects the unique ways in which you tend to make decisions. Your decision-making style and strengths and weaknesses are a vital aspect of the ways in which you govern.

A BUDDHIST POINT OF VIEW

The young son of a rich merchant was an enthusiastic Buddhist. He wanted to go home 'and use my wealth in doing good deeds.' Buddha asked him whether he could play the lute. 'When the strings of the lute were over-taut, did your lute give proper sounds?' 'No, Master.'

'When the strings of your lute were neither over-taut nor over-slack, the lute gave the proper sounds. Was it not so?' – 'It was so, Master.'

'Even so, an excess of zeal leads to self-exaltation, and a lack of zeal leads to indolence: have an evenness of zeal, master your powers in harmony. Let this be your aim.' (12)

A MUSLIM POINT OF VIEW

Riba has been defined as an 'increment, which, in an exchange or sale of a commodity, accumulates to the owner or lender without giving in return an equivalent countervalue or recompense to the other party.' (13)

'This unearned income without a corresponding productive economic activity or work, is the root of injustice and exploitation because it enriches the class of moneylenders and usurious who accumulate wealth by impoverishing those who are forced to borrow money or commodities from them for mere consumption of basic necessities.' (14)

Decison-making style

In the three case histories in Chapter 3, each person had a unique decision-making style as well as particular strengths and weaknesses. This style is a composite part of 'what you are' – some of it being innate, some of it shaped by experience and some a consequence of education and training.

Jane had a natural tendency to look for long-term trends. Her management training enhanced this strength. Lars was more action-oriented and good at completing challenging projects. Wang was able to scan large geographical areas, explore opportunities and then use his excellent sense of timing.

Each person has different decision-making patterns, and understanding these patterns is vitally important to corporate governance. Decision-making is one of the most important roles of any board member or investor. In my experience, the quality and timeliness of decision-making is critical to good governance. However, not many people stand back and ask whether the ways in which they are making critical choices are as effective as they could be.

Indeed, decision-making is often done automatically and unconsciously. How much more effective, then, is the person who is aware of how they are likely to make decisions and how their decision-making contributions are likely to affect the ways in which they govern.

A large part of self-governance is mastering the art of decision-making. When this is addressed, most board members are eager to improve their own and their group decision-making skills.

Decision-making is a complex process, involving all the facets discussed so far – core nature, life experience, values and beliefs plus acquired skills. One particular element has not yet been discussed, and that is the ability to get the best out of people. This will be explored further in Chapter 8, Guidelines for Interpersonal Governance.

There are many existing models that help to codify individual differences in decision-making style. In Chapter 5, which includes guidelines to effective decision-making, I shall refer to the Action Profiling System (15) as it seems to be the one most relevant to board-level decisions. The qualities that C.J. Jung (16) called 'functions' – that is, thinking, feeling, sensing and intuition – provide another simple map of the different ways in which individuals make decisions. Because I believe the intelligence spans more capabilities than intellect, I have used a wider notion of intelligence than the one normally associated with IQ tests.

Four key headings have been selected to define those areas relevant to the human side of corporate governance. They are the Physical, the Mental, the Emotional and the Spiritual. The qualities needed for many decisions in corporate governance are combinations of all four.

We probably use all these areas of intelligence when making decisions. However, the extent to which any one person is adept in each of these areas will affect the quality and nature of the ways in which they govern. We tend to be more developed in some than in others.

Physical intelligence

This is about the ability to get results in the physical world – not to entertain brilliant schemes, but to realize ideas in practice – to produce material products and services, and to make money. In terms of corporate governance, it involves making practical decisions concerning the material side of a corporation and ensuring that management delivers positive results. A whole group of intelligent and visionary people might dream up a wonderful business idea, but the one person that picks up the phone and makes it happen is the one showing physical intelligence.

Currently, business emphasizes this material side of life. Even when transactions seem highly abstract, for example commodities and futures trading, the desired end result is a lot of money in people's pockets. In terms of governing corporations, the current concern is to achieve the best financial results. This can involve mergers, acquisitions, takeovers and management buy-outs, selling companies or organic growth. All of these involve tangible, practical activities.

Even in the intellect-heavy world of the information technology business, there is a premium on getting results rather than simply having bright ideas, so there is less that needs to be said about this aspect of intelligence. It is firmly grounded in the business culture even though more academic types are sometimes less interested in getting things done (see box example).

WHEN PHYSICAL INTELLIGENCE IS OVERLOOKED

A CEO was frustrated because policies formulated at board level were taking up to two years to be implemented – if at all. The executive committee felt that he kept putting forward the same policies, despite them not being adopted. He could not work out what was going wrong.

Eventually, it became clear that there was no physical system for involving senior executives in the decision-making process. Managers could not relate to poorly communicated policies and could not see how to translate them into concrete activities. In some cases the directives were simply not feasible and they did not bother to implement them.

A system was created whereby senior managers were consulted prior to policies being formulated. They were asked to come up with practical implementation plans and a timescale which were then incorporated into the proposals. Once this was done, the problem disappeared.

However intelligent the policies may have been on other levels, they were lacking in what is here called 'physical intelligence.'

Mental intelligence

This is the intellectual capacity to understand concepts and to think logically.

It is closer to the common IQ idea of intelligence, but it is nevertheless broader in its scope. It includes the way we order our thoughts, create frames of reference and models, and argue for our ideas or beliefs. It involves absorbing and analyzing information, knowing how much detail is required, being able to get to the core of an issue and differentiating between what is important and what is not.

- How comfortable am I with the levels of information and understanding I need to have about the corporation I help to govern?
- Am I satisfied that I am sufficiently up to date with state-of-the-art thinking in the areas where I have to make decisions?
- What are the intellectual strengths that I bring to the board?
- Am I aware of how much I don't know?

Mental intelligence includes being able to make sense of figures or complex information and making them easily understood. Evaluating business plans and strategic thinking are both products of intellectual activity. It demands high levels of mental intelligence to assess how much information is required and how best to obtain it. Intellectual understanding is also required of technical details that are often outside the expertise of a board member. It also requires the ability to juggle mentally with the various factors involved in governance. For example, how to judge the quality of executive recommendations or the capacity of executives to actually deliver promised results. The ability to ask the right questions at the right time and to pinpoint problem areas requires a good mind and the ability to be objective, often in the face of considerable pressure.

Governance requires high levels of mental rigor and the ability to discriminate between things, and to think things through. This includes thinking through the human implications of various decisions. Effective governance occurs when there is a helicopter view of the big picture, and an understanding of how the parts fit together.

Multi-dimensional perspective is one of the most important aspects of governance, but one which is often lacking in boards because most board members have come from an operational rather than a conceptual background. Systems theory (17) has helped a great deal to enable directors to see how the whole process operates. People who have a broad sense of multi-dimensional perspective are few and far between. Often, their understanding is intuitive, abstract or spatial. It is usually complex.

One CEO used to use gestures to explain the interrelationships between the different elements of a situation. I once watched a president of a country mentally juggle the deployment of its national resources for 30 different services on being told that he had fewer resources than he anticipated. It took him three minutes to perform a complex process that could have taken a team of people at least a week. It is a rare gift to have this capacity and an even rarer gift to communicate it. Some people are intellectually brilliant. Most of us have to work hard at keeping our brains agile and alive.

At its best, this capacity enables us to discriminate between reality and fantasy, between what is important and trivial and between what serves humanity and what does not. For this, it needs to be used with wisdom, which combines understanding, compassion and experience, and an ability to know what is right.

Emotional intelligence

Emotional intelligence is a vital aspect of governance.

Without emotion we would act mechanically. Emotions make us feel that we are truly alive. The more in touch we are with them, the more effective we can be.

- How often do I really love myself and the people for whom I work?
- Am I sufficiently in touch with my own feelings to be able to deal effectively with them?
- Do I pay attention to what really makes me happy and fulfilled?
- Can I recognize and meet my deepest needs?

- How effective am I at dealing with difficult emotions?
- How sensitive am I to the effect of moral and emotional climate?
- How emotionally committed am I to what I am proposing or doing?

Without emotions it is impossible to be really committed to an idea or a decision. A great deal of human experience is emotional – love, hate, joy, anger, compassion, loneliness, security and fear. They are also a thermometer for gauging how things are working. Unhappiness about oneself, someone else or a situation is usually a warning sign that something is not right and needs to be addressed. If things are going well and there are healthy and constructive relationships between people who are emotionally committed to the same sense of purpose, then there is an almost tangible feel-good factor. Lack of emotional well-being can cause deep-seated unhappiness, resentment, frustration and emotional pain that drains energy and often causes worry, irritability and bad temper. Unproductive emotional conflict and tension between people is one of the major reasons for the breakdown of good governance and indeed, one of the causes of war. It is a symptom of emotional stuntedness.

There is increasing evidence that 'emotion is not a luxury: it is an expression of basic mechanisms of life regulation developed in evolution, and is indispensable for survival. It plays a critical role in virtually all aspects of learning, reasoning and creativity.' (18) Unfortunately, most of our educational systems do not help us to develop emotionally or, indeed, attach any importance to gaining emotional effectiveness. The world of feelings is extremely complex and most of us are mixtures of strengths and weaknesses.

The majority of board members are not consciously aware of their own emotions or the importance of this area. Neither are they aware that their emotional states have significant impact both on the wider group of people they govern, and on the people with whom they directly work. Out of four hundred board members, I have come across only twenty who were consciously aware of their emotions. These, commonly, had usually been through personal crises that forced them to look at their feelings. Some of them had then embarked on a long and often painful journey to learn to work more fruitfully with their emotions, often through therapy or other forms of self-development. Others were born into families where emotions were respected and had a natural interest in this dimension.

Emotional maturity involves the ability of each person to understand and manage their own feelings and to be sensitive and respond appropriately to the feelings of others. When spirit and purpose combine with a feeling of comradeship, shared determination to win through can achieve great things. The chances of success are increased, particularly if there is the intellectual understanding and practical resources and ability to make things happen.

THE STORY OF AN EMOTIONAL BULLY

Maurice was a refugee from a country where he had lost most of his relatives, including his father. He and his mother escaped to a new country where they set up a laundry business. Fifteen years later, there were chains of highly successful dry-cleaning shops all over the world. Share value was high. However, he never forgot the struggle to survive. 'Life is tough. You have to be tough to survive. Weak people do not survive long in this company.'

'The reason I have succeeded is that I fight to win. Kill or buy the competition and you will succeed.' His feelings about business rubbed off on the way in which he treated his directors. 'I am the boss. If you don't like what I say, you can leave.' His directors complained that 'he feels he has to dominate everything that moves and everybody within his sight.'

Many of his best people left as soon as they could, despite the fact that they were well paid. As serious was the fact that the company could not recruit good people from outside. 'They do not like the way people are treated here.'

At board level, his non-executives were too scared to challenge or argue with him. Some serious and costly mistakes were made that could have been avoided.

As indicated by the example, 'The story of an emotional bully,' emotional intelligence is closely connected with a person's values and beliefs. Emotional maturity not only enables a leader to treat his or her own people well, but often displays compassion and understanding for the plight of others. This affects the policies they recommend and the behaviors they will not allow.

Spiritual intelligence

STATE OF BEING

Values and beliefs are part of what I have called 'spiritual intelligence.' However, because they are so relevant to the human face of corporate governance, they were discussed as being important in their own right. The other aspect of 'spiritual intelligence,' that is, 'state of being,' is often related to emotional maturity. A positive and calm 'state of being' creates a presence and emotional environment that more easily enables constructive governance to take place. These are the sorts of questions that are often asked:

- Am I at peace with myself, my life and other people?
- Am I easily knocked off course?

- Do I suffer from impatience or high levels of anxiety or stress?
- Is the spirit in which I approach my work really how I would like it to be?

Some people refer to this as 'the higher state of being,' or 'a deeper understanding of the purpose of why we are here.' It impinges upon physical, mental and emotional intelligence. We make sense of the world, interpret it, and construct values to live by. However, we commit ourselves only to values that we emotionally feel good about. This in turn has a positive effect on physical health.

For the purposes of this book, one of the highest spiritual states is integrity, that is, the ability to be in touch with one's own truth, and to live by it. Someone may be technically brilliant, but if their integrity is not respected, and they lack a sense of personal authenticity, their ability to govern will be limited.

The ability to be truly reflective enables each person to gain perspective about themselves and their life and to check whether they are on the right track. After many years of training and discipline, mystics experience a profound feeling and understanding of oneness with life. The rest of us glimpse this understanding of reality for just a few moments in a lifetime. What we can do is to work towards achieving a spiritual state of peace, tranquillity and receptivity. This is the state where new insights are born and whence inspiration comes.

A positive state of being or attitude to life often makes the difference between success and failure. People who start with a negative attitude or expectation often behave in such a way that they actually attract the reactions that they most fear. The spirit in which a person governs has quite an effect on what happens. A restless spirit induces restlessness and discontent.

A common problem is that leaders worry a great deal. They can't sleep and spend a lot of energy being distracted because they can't stop thinking about a challenge or difficulty. 'This is very tiring. I wish I could do something about it.' The advice that follows sounds easy, but is in fact extremely difficult to put into practice.

When you have a problem or have done something wrong, do what you can to deal with it. After that, be still. Cease the endless circling of the over-scrupulous mind. Then move on. (19)

When the spirit of a person is joyful and positive, it is usually because they are acting in line with their values in a way that is personally rewarding to them. This can be so in spite of extremely adverse circumstances.

It often requires considerable effort to discipline oneself to develop the capacity to control one's state of being at will. It usually requires considerable training, patience and the ability to reflect. A common refrain is 'this is not the top of my list of priorities, I genuinely do not have enough time for this'.

EXAMPLE OF STATE OF BEING OF A COMPANY

In South Africa, one company increased its productivity by 15 per cent for three years running, in spite of a lack of general and managerial skills. One main board member said that it was 'partly due to good strategy. But the most important thing is that the place is alive and humming.' What was the secret?

The CEO had asked his senior managers how they could 'wake things up.' After consultation, an African tradition of acknowledging success through a praise singer was adopted. 'When a group of people does exceptionally well, they invite people, including the main board, to celebrate. A praise singer comes and wholeheartedly supports that group. You can't keep any of them down,' the CEO said, 'it is so much fun that personally I can't wait to get to work.'

Another is 'if I do too much soul searching, I might discover some things about myself that I don't like,' or 'once I start navel gazing, I will lose my practical edge.' However, for those for whom achieving a consistently good state of mind is important, many find the effort well worth it.

The last few pages have described a number of components that make up the character of an individual and which are considered to be important ingredients for effective self-governance. They make up some aspects of 'who you are,' and affect 'what you do.' During the course of this chapter, a number of examples have been given in terms of what people do in respect to some of the different aspects mentioned. However, in terms of how an individual contributes in practice to corporate governance, three main areas of activity deserve particular attention. These are the sharing of knowledge and expertise, the type of governance a person is involved in, and the ways in which any business leader relates to other human beings.

INTUITION

There is another state of being that is included in this section – being open to intuition. It is one of the most inspirational and effective forms of decision-making. It is one that many board members use, consciously or unconsciously. It is defined as 'spiritual perception or immediate knowledge.' Hawthorne in 1851 described it as 'some truths so plain and evident that they do not need any rationalization process.' Many people will identify with having a 'miraculous intuition for what ought to be done at just the right time for action.' (20) Intuition is not tangible and therefore it is difficult to determine how accurate a hunch is. As one chairman said,

'Ninety per cent of the time I'm right, but I do need to back it up with facts or I could make some terrible mistakes.' Most people, however, pay a price for ignoring their intuition as it is often an early warning about something that needs attention.

WHAT YOU DO

Share knowledge and expertise

It is hoped that by now it is clear that each individual brings much more to corporate governance than is usually written in their CV. They bring a wealth of unique experience and wisdom over and above technical competence, although technical competence is a prerequisite. This includes the ability to understand and carry out legal and fiduciary duties, to understand finance, the industry and markets, knowledge of the actual businesses being governed, and what it takes for any one company or group to deliver results. It also involves a certain level of interactive and communication skills. This will be covered in greater detail later in the book.

The three characters in Chapter 3 all had the necessary technical experience, skills and knowledge. In addition, Jane was extremely good with people and encouraged her directors to develop and excel. She was a good mentor to others. She was also good at avoiding confrontation and establishing a collegiate environment on her board that encouraged good debate. She also inspired people to keep ahead of the market by her long-term vision and understanding of trends. Lars was excellent at getting things done and running complex tangible projects. He motivated people to deliver consistently good results. His enthusiasm kept people going in difficult circumstances. In addition, he was an excellent public speaker, and championed the causes he believed in. Wang found organizations and people with potential, invested in them and let them have access to his knowledge. He was a master networker, and had relevant contacts all over the world. He was a brilliant opportunist and deal-maker. One of his strengths was seeing what was going on at any one time on a global scale, and then making sure that he was in the right place with the right people at the right time. He created an organic group of companies that was well able to flex, group and regroup according to short-term market demands.

The unique ways in which each person makes decisions is a combination of their natural tendencies and what they have learned. The way each person applies their knowledge of the decision-making process contributes to the practice of governance. Jane will make sure that the background of any challenge is thoroughly understood. She is also good at spelling out both positive and negative implications of any major decision. Lars is good at insisting that a case for a decision is thoroughly argued, and that any decision is feasible. Wang is good at exploring and generating options, creating a sense of urgency and asking questions about the timing of opportunities.

Participate in governance

Decision-making is an essential element of governance. However, this needs to be appropriate to both the type and level of governance each person is involved in. Type of governance has to do with whether the predominant role of a board or committee is to lead, direct, monitor and control, supervise the executive or manage external relationships. It also depends on the kind of pressure that major investors and shareholders might apply. The level of governance depends on whether a person is best suited to working with boards and committees, intergroup work or operating on a more systemic level. The levels at which they mainly operate will affect the ways in which they impact on others, whether it be to influence individuals, groups of people, organizations or communities.

Relate to others

The way each person impacts on others is directly affected by the ways in which they relate to others. This depends both on their interpersonal skills and the values and beliefs they bring with them concerning people. Jane led in a highly participative manner because she believed in involving as many people as possible. During meetings, Lars was more directive. He believed in the value of clear leadership and then delegating activities to autonomous but accountable people. Wang was more invisible. He preferred to let people get on with it.

The attitudes each person has to power and authority will also influence how they participate as a member of a group and how they are likely to contribute. It also depends on their natural tendencies. Jane preferred to work more on a one-on-one basis. Lars enjoyed large groups, and Wang preferred to observe and intervene only when necessary.

The way each person relates to his or her colleagues is also part of this. If someone is either suspicious or trusting of others, this will affect the way others react to them. Their views on themselves as leaders will also influence the kind of cultural environment they will promote, be it democratic, autocratic, anarchistic or a confused mixture of all three.

SUMMARY AND CONCLUSIONS

In this chapter, a model was provided to enable each person to gain greater self-knowledge about themselves and to understand more about where others are coming from. The diamond was the metaphor used, and the expanded diagram in Figure 4.2 shows in more detail the component parts.

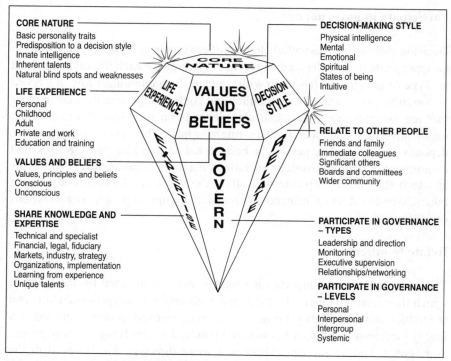

CORE NATURE
Basic personality traits
Predisposition to a decision style
Innate intelligence
Inherent talents
Natural blind spots and weaknesses

LIFE EXPERIENCE
Personal
Childhood
Adult
Private and work
Education and training

VALUES AND BELIEFS
Values, principles and beliefs
Conscious
Unconscious

SHARE KNOWLEDGE AND EXPERTISE
Technical and specialist
Financial, legal, fiduciary
Markets, industry, strategy
Organizations, implementation
Learning from experience
Unique talents

DECISION-MAKING STYLE
Physical intelligence
Mental
Emotional
Spiritual
States of being
Intuitive

RELATE TO OTHER PEOPLE
Friends and family
Immediate colleagues
Significant others
Boards and committees
Wider community

PARTICIPATE IN GOVERNANCE – TYPES
Leadership and direction
Monitoring
Executive supervision
Relationships/networking

PARTICIPATE IN GOVERNANCE – LEVELS
Personal
Interpersonal
Intergroup
Systemic

Fig. 4.2   Components of self-governance

How effective someone is depends on the extent to which they learn to manage different aspects of themselves, and themselves as a whole person. This will influence the way in which each person manages or mismanages their priorities, time and energy. Who someone is and what they do fundamentally affects the difference they make to other people and whether the way they lead their lives really meets their needs and those of others.

How each person responds to circumstances also depends on how well developed they are in terms of how they function mentally, emotionally, spiritually and physically. This in turn influences the ways in which each person makes decisions and major choices about their own lives and those of others.

This chapter was about definitions. The next suggests some guidelines for self-governance for those committed to improving the ways in which they govern.

Notes
1. Soros, G. (1998) *The Crisis of Global Capitalism*, Little, Brown and Co.
2. Gazzaniga, Professor M.S. (2000) *Automatic Brains/Interpretive Minds*, The Royal Institution of Great Britain, Lecture List, Page 6.
3. Hopkins, G. M. (1961) *Inversnaid*, Penguin.
4. Ridley, M. (1996) *The Origins of Virtue*, Penguin Books.
5. Damasio, Professor A., Royal Institution lectures, ibid. p. 2 and Damasio, Professor A., (2000) *The Feeling of What Happens*, Heineman.
6. Carneson-McGregor, L. (1994) *Homage To Hope*, Aspen Books, 1994.
7. The Holy Bible (1980) *Matthew XXV*, 14:30, Collins.
8. The Holy Bible (1980) 1 *Corinthians XIII*, Collins.
9. Director of Investment Company.
10. McGregor, L. (1975), *Learning through Drama*, Heineman.
11. Demb, A. and Neubauer, F. (1992) *The Corporate Board. Confronting the Paradoxes*, Oxford University Press.
12. Translated by Mascaro, J. (1973) The Dhammapada, The Bhagavad Gita 616, Penguin.
13. Al-Mabsut, vol. 7. P. 109 – http://www.iol.ie/ – afifi/articles/riba.htm
14. *Riba and Investing in Shares*, http://www.iol.ie/ – afifi/articles/riba.htm
15. Ramsden, P. and Zacharias, J. (1993) *Action Profiling – Generating competitive edge through realizing management potential*, Gower.
16. Jung, C.J. (1966), *The Spirit in Man, Art and Literature*. (The Collected Works of C.J. Jung, Vol. 15), Princeton University Press.
17. Goleman, D. (1996) *Emotional Intelligence*, Bloomsbury, and Witkin, R.W. (1974) *The Intelligence of Feeling*, Heineman.
18. Damasio, Professor A. (2000) *The Neurology of Emotion*, Royal Institution of Great Britain Lecture List (p. 2).
19. Vann, G. (1985) *The Divine Pity*, Fount Paperbacks.
20. The Compact Edition of the *Oxford English Dictionary*, 1979, Oxford University Press.

5   Guidelines for Self-Governance

How do I know how well I am doing?
How can I improve what I do?

▶ ▶ ▶   **EXECUTIVE SUMMARY**   ◀ ◀ ◀

This chapter offers practical ways for improving self-governance. Six basic principles are offered as being basic to effective governance. The chapter suggests that greater self-awareness leads to greater ability to manage one's strengths and weaknesses. This includes greater consciousness of the unique contributions and talent each person has to offer. In addition, because corporate governance is based on values and beliefs, it is useful on a regular basis to re-evaluate one's own values and principles.

Self-governance is a journey that is different for each person. Each person is unique, their situation in time and place is unique, and their specific areas of interest or what each person feels they want to learn is a matter of individual choice. Consequently, this chapter does not attempt to map that journey for everyone; it simply suggests ways to begin it – because beginning this journey is often the most difficult part.

Two sorts of guidelines are given. Six basic principles for self-governance are described, together with suggestions for putting them into practice. In Chapter 8 there is an outline of the formal governance roles in an organization. These formal roles clearly impinge upon the individual's requirements for self-governance, so the reader is advised to check the expectations and obligations of their formal role before following the practical guidelines in this chapter in greater depth.

The six principles sound obvious, but are not always easy to put into practice. They apply to anyone involved in governance, regardless of their role. The principles are as follows:

- Know yourself.
- Give of your best.
- Manage your weaknesses and blind spots.

- Be ruthless with the way you spend your time.
- Take care of your health.
- Deal promptly with difficult relationships.

THE APPROACH TO SELF-GOVERNANCE

The reader can approach this chapter in a number of ways. Basically, it is possible to assimilate the general principles and get value out of them. It is also possible to select one or two key areas and work on those, or it is possible to begin with a broad exploration of the self and then focus on those areas most needing development. The reader may decide to continue the work at a deeper level. In this case, it is recommended that they obtain expert advice or assistance.

Although I have tried as far as possible to supply a do-it-yourself set of recommendations so that the reader can explore as far as possible on their own, it is not easy to go further without some form of tutor or mentor. One of the best ways of gaining self-awareness is by holding a mirror up to oneself to gain constructive feedback, not only from colleagues but also from objective non-partisan outsiders.

My extensive experience of helping those involved in corporate governance has confirmed that there are no easy answers. Each person is special and has specific issues and areas that they want to develop. Each person also learns best in different ways. So the reader is invited to take responsibility and select what they choose to learn and how they can best progress that learning.

A prince must, therefore, always seek advice. But he must do so when he wants to, not when others want him to ....

Here is an infallible rule: a prince who is not himself wise cannot be well advised. (1)

To get the best out of this chapter, it is recommended that you use the questions posed later in the chapter to decide which improvements would make the greatest difference both to your work and to your life, and set these as priorities. Certain things take only a minute to accomplish; others may take years of training and dedication. Self-development is a life-long journey. Personal governance is the starting point because the relationship that any person has with their self will reflect on their relationship with others and, ultimately, on how they govern.

To take self-governance seriously demands commitment and the ability to reflect. It also takes time. In this chapter I have endeavored to take into account that those involved in corporate governance have very limited time. However, even a small initial exploration on the lines I suggest can have a considerable effect, and that in turn can encourage further effort. So the reader is invited to take at least one area and work with it.

In preparing these guidelines I have also concentrated on those issues most relevant to corporate governance from my personal experience of working with both non-executive and executive board members. The six principles of self-governance have been distilled from that experience.

This chapter focuses particularly on positive attributes. Even when we address weaknesses and blind spots the emphasis is on making the best of what we have got. There will be a separate section on how to deal with the dark side of human nature in Chapter 9 which explores the shadow side of corporate governance.

SUMMARY OF THE PRINCIPLES OF SELF-GOVERNANCE

This summary provides a quick overview of the principles, which will be discussed later in the chapter. If you wish to work on self-governance, it is suggested that you give yourself a rating out of ten for each of the six principles. This will serve as a guide and help you to focus on areas needing most attention.

1. **Know yourself**
    How well do you think you know yourself?
    (Rating out of 10: ......).

    - Be aware of your strengths, decision-making style, unique talents and expertise.
    - Be aware of what you need in order to give of your best and share this with others so that they can help you to persist.
    - Be true to yourself at all times. Only say yes if it feels good and right to say yes.
    - Regularly ask people you trust to give you constructive feedback about yourself.

2. **Give of your best**
    To what extent do you feel you really give your best?
    (Rating out of 10: ......).

    - Focus on what you do best. Develop your strengths.
    - Ascertain which part of the life cycle and level of business you can most contribute to, whether it be initiating new businesses, maintaining and growing existing ones, restructuring and so on.

- Make sure you are at the right place with the right people at the right time.
- Learn to say no when you need to.

3. **Manage your weaknesses and blind spots**
   How well do you think you recognize and manage your weaknesses and blind spots?
   (Rating out of 10: ......).

   - Let go of the need to be perfect. Be compassionate and honest about your weaknesses.
   - Become aware of the effects your weaknesses and blind spots can have on yourself, the way you govern and on other people.
   - Learn to limit any damage these may cause other people as quickly as possible.
   - If it is difficult to overcome some of your weaknesses, ask people you trust to help compensate or mitigate them.

4. **Be ruthless with the way you spend your time**
   How satisfied are you with the way you spend your time?
   (Rating out of 10: ......).

   - Save your time for what you really do well.
   - As to the things you don't do well, find others who will do them better.
   - Follow your heart as well as your head.
   - Learn to use and respect your natural rhythms and sense of timing.

5. **Take care of your health**
   Are you as healthy as you would like to be?
   (Rating out of 10: ......).

   - Make good health a priority.
   - Ask family and friends to warn you if they see signs of burnout.
   - Check that your real needs are being met in healthy and effective ways.
   - Seek the right balance between private and work life, giving and receiving and being active and reflective.

6. **Deal promptly with difficult relationships**
   How effective are you at dealing with difficult relationships or people?
   (Rating out of 10: ......).

   - If you have problems with other people, make sure you sort them out as soon as possible. Do not wait until the problems get worse.

- The same applies if others have problems with you. Ascertain what the problem is and deal with it.
- Take your intuitions seriously. Ignore them at your peril.
- If someone is not performing, deal with them promptly. If they are unlikely to deliver, let them go as soon as possible.

1. KNOW YOURSELF

Self-knowledge is the starting point. Unless each person really understands what they have to offer and what their limitations are, governing will be a hit-and-miss affair. Unless someone knows how they are likely to affect others, it is difficult to know how effective they can be. One's relationship to oneself and the way values and beliefs are lived out will fundamentally influence relationships with others.

In the last chapter, a model was suggested as a basis for self-understanding. If you rated yourself out of ten in the above summary, it could help you get some idea how much you know about yourself, but the questionnaire later in this chapter takes the enquiry much further. It also outlines a program.

Because it is not always easy to divide what is innate and what is acquired, the questions are intended to help you decide what you can live with, what you can change, and what you are unlikely to change. I have found that the closer a person is to who they feel they really are, the more effective they are. The purpose, therefore, is not to try to change your core nature, but rather to enhance the positive aspects.

The questionnaire is designed first and foremost for individual exploration, but it is not a substitute for having constructive feedback from others as well. A good mentor is likely to understand these aspects, even though they may use a different framework or language to describe them. They should then be able to challenge you, or enable you to find out what is most important at any one time.

2. GIVE OF YOUR BEST

Every person has unique talents – ways of making decisions and contributing to improving the quality of governance. One of the best ways, therefore, of adding value is in the main to focus on and develop what you do best and what you really enjoy doing. This is what most highly successful people do, often subconsciously. If you are not sure what your real talents are and whether you are in an environment where you can do your best, a starting point is the questionnaire later in this chapter. However, it is well worth the time, effort and money to find a good specialist to help you.

Consciously choose those circumstances and people with whom you feel you can make a real difference and where you will be happiest. As one chairman put it, 'life is too short to stay any time in an unpleasant environment.' If your level of fun is low and your level of misery is high, it may be time for you to move on.

It might also be useful to know that different people do better at different points in the life cycle of an organization. Those who shine during the start-up phase of companies are seldom interested in growth, consolidation or maintenance and so on. Trying to govern or run a company that does not fit your profile is like fitting a round peg into a square hole.

3. MANAGE YOUR WEAKNESSES AND BLIND SPOTS

In a highly competitive climate, there is enormous pressure to be perfect, so people tend to deny that they have any weaknesses or blind spots. This is not possible. Everyone has them.

This could include gaps in decision-making, or strength in one area being balanced by lack of development in another. For example, in the business world many people are under-developed in terms of emotional and spiritual intelligence – skills not well addressed in our educational systems. Others may be good at tactics but less good at long-term strategy and vice versa.

There are two good ways of testing for weaknesses and blind spots. The first is to list at least five different mistakes that you have made over the years. Try to identify the causes and you will probably find some of the same weaknesses recur. The second way is to ask others what they feel your weaknesses to be. Secretaries and personal assistants often know a great deal, followed by the company secretary and then colleagues and members of your family.

Finally, everyone has a potentially fatal flaw: a blindness that can lead to a fall, or a destructive impulse rooted in the dark side of one's nature. A need to be liked at the expense of vigilance is an example of such a flaw. In the stories in Chapter 3, Jane avoided confrontation. Lars was too domineering in the wrong circumstances. Wang's invisibility might leave others feeling abandoned in times of crisis. The main emphasis in this chapter is on the more positive aspects of self-governance, and the possibility of fatal flaws is discussed in greater detail in Chapter 9 on the Shadow Side of Corporate Governance. In that chapter, there are some recommendations on how to deal with such important flaws. If, however, certain problems are deep-seated and cause recurrent difficulties, it is better to consult a psychotherapist or behavioral psychologist.

Otherwise, the best general approach to overcoming weaknesses is to create an environment where human failings are acknowledged, and where people support each other to mitigate or avoid each others' excesses. Luckily, the other person's failings are seldom identical to your own.

4. BE RUTHLESS WITH THE WAY YOU SPEND YOUR TIME

One eminent investor, who was a professional non-executive chairman and non-executive director of a number of companies, decided to resign from most of them. His problem was lack of time. In fact, the most common complaint of directors is that there is not enough time for them to make reasonable decisions, and they are not given information sufficiently in advance of meetings. 'We are really at the mercy of the CEO if he doesn't give us enough time to digest the material or to challenge his thinking.'

Lack of time can often be attributed to two sources. The first is an inability to say no. The second is trying to do everything, as well as wanting to be where the action is. In the initial summary, four main suggestions were given to help this: focus your time on what you do well; delegate to others; follow your heart in terms of what you really want to do; and learn to respect and use your natural rhythms.

Each person has natural rhythms and ways of choreographing their energy in a way that best suits them. Going against these can cause stress, particularly for others. For example, the CEO of a major computer company always arrived late and always ran late meetings. He was not popular with his directors. Through creative techniques of self-exploration he discovered that he did his best thinking when things were open-ended. What he needed to do was to take this into account and create ways in which he could be both flexible and meet the needs of his colleagues. Compromises were reached and everyone was happy.

As one director put it, the most important thing to remember about time is that 'life is not a dress rehearsal.' Time wasted can never be regained. Some people find planning essential. Others are effective because they choose courses of action where they know the quality of experience that they have will be fulfilling. These people are seldom tired or drained. It is all a matter of being highly selective rather than rushing around 'being busy.'

5. TAKE CARE OF YOUR HEALTH

Healthy chairmen and directors are not necessarily the best governors. Throughout history there have been cases of successful, even brilliant, business or political leaders who have been either physically or mentally ill. Maybe they are driven to extremes of success because of their afflictions. However, as far as I know, no one has determined whether they would have been just as good had they been healthier.

Ill health among lesser mortals, however, does definitely reduce productivity. Falling asleep at board meetings is not only a symptom of boredom; it is often a symptom of stress and fatigue. In dysfunctional organizations or boards, where members are deeply unhappy and frustrated, there are high incidences of anxiety, stress and paranoia as well as insomnia, backache and other physical ailments. These drain energy and distract people from doing their best. What is more, stressed board members take their problems home, and that leads to strained domestic relationships and even greater pressure as a vicious cycle develops. It is not uncommon to find a higher rate of illnesses among family members where the spouses are highly stressed.

It is clearly advisable to maintain a healthy lifestyle as far as possible. The trend is towards healthier living. More people in leadership positions drink and eat less nowadays, and they eat healthier foods. They also exercise more. However, for those like me who enjoy their food and chocolate, it can be a constant battle.

Everyone has different ways of staying healthy. The following suggestions are just a few of the tips I have picked up over the years.

- Make health a life priority. Have regular check-ups. Find exercise that you enjoy. If you have limited time or self-discipline, either buy exercise machines or have a regular coach who can make sure you take exercise.
- Massage and relaxation techniques are useful for stress and jet lag.
- Ask your family and friends to point out when you are over-tired. Often you are the last person to know when the best time might be to take a break.
- If you find you are practicing addictive behavior, such as overworking, over-eating or drinking or worrying too much, ask yourself whether you have real needs that are not being met. If this is so, find healthy ways of fulfilling these needs. The more you neglect real needs, the worse they become.
- Try to find the right balance for yourself between work and recreation, private and social life. To give and not to receive, or to work without rest, often produces stale and boring results. Make sure you have enough play and fun in your life.
- Continued learning is said to keep one healthy and alert. Figure 5.1 can be used to check whether there are any areas that you might like to develop either in work or in life. The theory is that the wider the range of experiences, the more there is to offer.

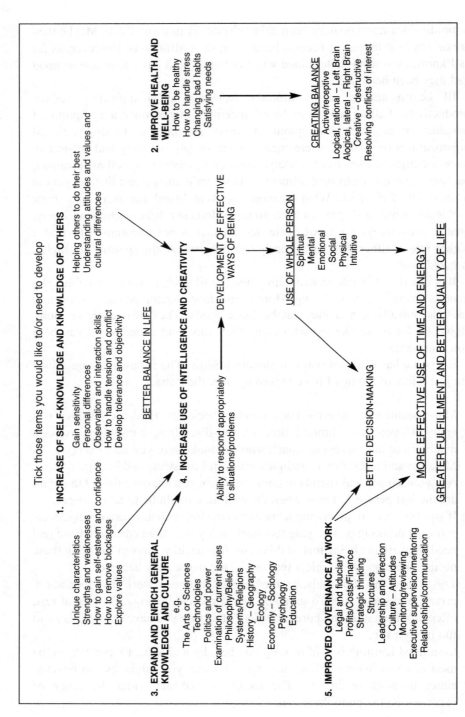

Fig. 5.1   Aspects of self-development

6. DEAL PROMPTLY WITH DIFFICULT RELATIONSHIPS

In my experience, the biggest difficulties that any board member suffers are the effects of unresolved relationship problems that fester, deteriorate and then escalate. The effects can be disastrous. People become angry and stop co-operating. They can withhold crucial information, encourage factionalism and cause many other difficulties. The usual pattern is for people to avoid areas of interpersonal difficulties. People are too polite and dislike unpleasantness. However, in almost all cases, the sooner these things can be resolved, the better.

Some suggestions were mentioned in the summary earlier. These included dealing with difficulties as early as possible, being alive to the possibility that you may be causing the difficulty, trusting your intuition and dealing promptly with people who are not performing well. In addition, the following guidelines may be of use.

- If you sense a problem, check it out first with the other person. You may not be right.
- Do not pass judgment unless you are absolutely sure you understand the situation.
- Always initially assume good intent.
- If you do something wrong, admit your mistake and apologize.
- If you have trouble with someone, describe his or her behavior accurately and without exaggeration. Then express the way you feel about it and the effects the behavior has had. Then specify what you want done differently, and how you may be able to support the change. It is more likely that you will know of better solutions than the person concerned. Spell out positive consequences if the person changes his or her behavior or the negative consequences if they do not. If you intend to act, then it is important that you commit to carry out what you say you will. (2)
- If someone has problems with you, ask them what they need for the problem to be resolved.
- Avoid taking responsibility for another person's feelings. An apparently conscientious decision like 'I won't say that in case it hurts his feelings' can lead to major problems in the long run.

For those who want to go more deeply into this subject, the chapters on Guidelines for Interpersonal Governance (Chapter 8) and The Shadow Side of Corporate Governance (Chapter 9) will deal with some of these aspects in more detail.

QUESTIONNAIRE FOR SELF-GOVERNANCE

The best way to tackle these questions is to try to answer them quickly and without too much thought. You can then check whether they ring true or not. There may be some areas where you may not know the answers. If this is the case, or if you want to explore some aspects in more depth, then suggestions are made for you to take them further. This is first and foremost a private exercise, so I suggest that you be as honest with yourself as possible.

Although these questions should be addressed initially on your own, it is thoroughly recommended that, if you wish to proceed further, you discuss them with a mentor or counselor, or work together with a friend or colleague. The more constructive feedback you get from well-meaning people who know you well, the greater the level of self-awareness.

Remember that there are two reasons for gaining better self-knowledge. The first is to decide which areas you would like to improve. The second is to use your reflections as a basis for sharing with other people your preferred ways of thinking and working. In this way they can take into account the best ways of working with you.

1. CORE NATURE

(a) Briefly describe those characteristics that you think are core to your nature. Any of these may be relevant: basic personality traits, predisposition to ways of making decisions; innate intelligence; natural talents; certain blind spots and weaknesses. For example, a talent for getting to the essence of a problem or a weakness in not being good with details.

(b) Which of these do you feel has a positive or negative effect? For example, the ability to move at speed as a positive attribute, or impatience as a negative quality.
  1. Positive effects
  2. Negative effects

(c) List what you consider to be your main talents and areas where you feel you have outstanding ability.

2. LIFE EXPERIENCE

(a) Briefly describe how your parents might have influenced you.

(b) Describe at least three key events in your childhood that affected you. How? What did you learn from them?

(c) Describe at least three key events or people that influenced you as an adult. How did they influence you? What did you learn?

(d) If a young person were to ask you what you learned from your experiences and what you would like to pass on to them, what would you say?

3. VALUES AND BELIEFS

POSITIVE VALUES AND BELIEFS

(a) List your key values, for example, honesty, loyalty, winning.

(b) List the key principles by which you try to live. It is important to tell the truth, however difficult it may be.

(c) To what extent do you think that the people with whom you work have the same values?

(d) With a top score of ten, rate the extent to which you feel you are living according to your values and principles.

(e) In terms of the human side of corporate governance, what is the legacy you would like to leave behind? What is the human face you think you present?

(f) *You may wish to explore more deeply what your current values and beliefs are. You may also wish to reaffirm your values and beliefs. If so, the following exercises may be useful:*

    1. Briefly describe a difficult issue that you resolved in a satisfactory way. List what you think were the values and beliefs behind it.

    2. Think of the last occasion when you were annoyed or angry at something or someone. Ask yourself which of your values and principles were undermined.

NEGATIVE BELIEFS

Negative beliefs are often pessimistic, unconscious beliefs that influence your responses to situations. For example, if you believe that all people are untrustworthy, you are likely to be suspicious of anyone regardless of whether they are trustworthy or not. You may also have negative beliefs about yourself. For example, you may think that as a person you are basically unworthy or a fraud. You will, therefore, constantly try to prove that you are otherwise.

(a) Catch yourself next time you feel unhappy with yourself. Ask yourself, 'What is the feeling behind this? Why am I feeling this? What is the belief behind this? Is it true?'
   If it is not true, you have understood an unconscious pattern that no longer serves you and you can let it go.

(b) If you feel judgmental or angry with someone, ask, 'What is the cause of my judgment or anger? Is it specific to the situation, or is it something I believe in general? Is my assumption really correct about that person or group of people? If not, is it time to let it go?'

▶

4. DECISION-MAKING STYLE

(a) Think about a specific decision involving others that you are about to make, or that you have made. Briefly describe how you like to go about making decisions. For example, do you need information or options first, or do you prefer to have a gut feel for what you want and then seize opportunities, or a combination of different ways.

(b) Tick which types of intelligence you feel you are strong in. Put a cross next to the ones that you feel are lacking.

Physical – ability to organize things in tangible and concrete ways. You are good at knowing what is needed to make things happen.

Mental – ability to see things in terms of concepts or abstract theories. Also, ability to think logically and rationally. Good with theories.

Emotional – good at being in touch with your own and sensitive and responsive to other people's feelings. Ability to express your own emotions and to deal easily with emotional issues.

Spiritual – you have a clear ethical sense of what is right or wrong, and human values. You are content about your spiritual state of well-being.

Spiritual/intuitive. You feel you operate intuitively much of the time.

(c) *Modes of thinking.* Figure 5.2 describes different modes of thinking and should indicate where your natural decision-making preferences are likely to be. (3) Indicate whether you have a strong preference to think in certain ways (H), or whether you have low areas of motivation or blind spots in certain areas (L). Also indicate in each thought mode whether you think you prefer to share your thinking or go private. If you feel you may do both at times, indicate with an S or P. If you are not sure, it might be worth having your decision-profile done by an expert.

5. SHARING KNOWLEDGE AND EXPERTISE

(a) To discover your areas of outstanding talent, a good way is to list three or four events in your life that were high spots and where you felt a significant sense of achievement.

(b) Describe each and then work out what qualities you applied in each event. When you have done three or four, you will find certain qualities in common. These will point to what your particular talents are.

(c) Do the same in terms of the environment and circumstances, and you will have a clue about what are the best environments in which you excel.

(d) List your more general areas of knowledge and expertise.

In terms of your particular specialist knowledge and expertise.

In terms of your role on the board or boards of which you are a member.

To get a feel for your modes of thinking and sharing preferences

Tick appropriate boxes against each thought mode

Motivations: High (H); Medium (M); Low (L)   Preferences for sharing (S) or Private thinking (P)

Action oriented
Making the effort to complete the chosen task

Perspective/conceptual
Viewing the chosen task from different perspectives

Attending to information and options

Investigating – What already exists

	Motivation		
	H	M	L
Probing			
• Probing for information	☐	☐	☐
• Making distinctions	☐	☐	☐
Classifying			
• Making connections	☐	☐	☐
• Grouping data	☐	☐	☐

Preferred interaction
Sharing   ☐
Private   ☐

Exploring – What is possible

	Motivation		
	H	M	L
Encompassing			
• Gathering possibilities from different angles	☐	☐	☐
Generating			
• Generating alternatives	☐	☐	☐

Preferred interaction
Sharing   ☐
Private   ☐

Intending to take action

Determining – What needs to be done

	Motivation		
	H	M	L
Pressure			
• Building a firm case	☐	☐	☐
• Applying or resisting pressure	☐	☐	☐
Persisting			
• Gently persisting	☐	☐	☐
• Bending without giving way	☐	☐	☐

Preferred interaction
Sharing   ☐
Private   ☐

Evaluating – What is important

	Motivation		
	H	M	L
Prioritizing			
• Establishing relevant issues			
• Ranking issues	☐	☐	☐
Comparing			
• Comparing options			
• Weighing pros and cons	☐	☐	☐

Preferred interaction
Sharing   ☐
Private   ☐

Committing to action

Timing – The pace of action

	Motivation		
	H	M	L
Opportunity seizing			
• Speeding up	☐	☐	☐
• Seizing opportunities	☐	☐	☐
Pacing			
• Choosing the right time to act	☐	☐	☐
• Taking time	☐	☐	☐

Preferred interaction
Sharing   ☐
Private   ☐

Anticipating – The progress of action

	Motivation		
	H	M	L
Reviewing			
• Planning the stages of action			
• Measuring progress	☐	☐	☐
Goal setting			
• Setting goals/visions			
• Seeing implications	☐	☐	☐

Preferred interaction
Sharing   ☐
Private   ☐

Fig. 5.2   Modes of thinking

In terms of your contribution to decision-making in the boards or committees of which you are a member.

(e) With a possible score of ten for each of these, rate how satisfied you are that you are contributing your best. If any of these is under six, you are operating significantly below your potential.

6. PARTICIPATION IN GOVERNANCE

(a) Which levels do you feel you contribute most to?
   ☐ Improving the system.
   ☐ Representing groups or influencing relationships between formal groups.
   ☐ Running or contributing to interpersonal meetings.
   ☐ One-on-one working relationships.
   ☐ Working on your own to come up with solutions to difficult challenges.
   ☐ A combination of all of these.
(b) Are you best at any of the following?
   ☐ Giving leadership and direction.
   ☐ Strategic thinking.
   ☐ Seizing opportunities.
   ☐ Mergers, acquisitions, joint ventures and so on.
   ☐ Negotiations. Doing deals.
   ☐ Mediation.
   ☐ Structuring and restructuring.
   ☐ Providing vision.
   ☐ Monitoring.
   ☐ Overseeing the executive.
   ☐ Executive activities.
   ☐ Developing and maintaining important relationships.
   ☐ Public relations. Communication. Working with the media and so on.
   ☐ Any others.

7. RELATING TO PEOPLE

(a) Briefly describe your leadership style and your leadership strengths. How do you prefer to get the best out of people?
(b) Briefly describe how you operate as a chairman/woman.
(c) How would you like to be a better chairman/woman?
(d) Briefly describe how good you are at participating in decision-making and influencing decisions.
(e) How would you like to improve your influencing skills?

(f) Out of ten, how would you rate your capacity to listen?

(g) Out of ten, rate how good you feel you are at dealing with interpersonal conflict, tensions and hostility.

(h) How do you prefer to relate to:
   A. The main board?
   B. The executive?
   C. Other significant groups, for example major investors and other shareholders?

8. MAIN STRENGTHS AND WEAKNESSES

Use the answers you have come up with so far to summarize your main strengths and weaknesses.

(a) List what you consider to be your main strengths.

(b) List what you consider to be your main weaknesses.

(c) List what others consider to be your main strengths.

(d) List what others consider to be your main weaknesses.

9. PREFERRED WAYS OF WORKING

(a) Briefly describe your preferred ways of working.

(b) Describe what you personally need in order to be at your best.

(c) What is the type and amount of information you require to make good decisions?

(d) Describe your least preferred ways of working, such as the types of people, behavior or situations you would prefer to avoid, for example bullies or working to impossible deadlines.

10. LIST WHAT YOU CONSIDER TO BE YOUR MAIN NEEDS

(a) As a human being.

(b) As a person involved in corporate governance.

11. AREAS FOR SELF-DEVELOPMENT

(a) List two areas that, if they were improved, would make a significant difference.

(b) Bearing these in mind, list not more than four key areas for self-development that you are prepared to commit to.

12. NEXT STEPS

List your next steps and your time-scale, including with whom, how, what, where and when you wish to work on your key aspects of self-development.

SUMMARY AND CONCLUSIONS

In this section, guidelines were given for self-governance. It was suggested that if the reader followed all or some of the principles advocated above, the unique ways in which they govern would be improved.

Six principles were offered. The first was the importance of self-awareness. The second was a suggestion that if you focused on giving your best, you would be more effective. On the other hand, it is also important to recognize and manage your weaknesses and blind spots. Another recommendation is to be ruthless with the way in which you spend your time. To some extent this is related to taking care of your health. Finally, it was suggested that difficult relationships were dealt with speedily and promptly.

All of these principles apply to anyone's self-governance, regardless of the formal role or roles they might play. However, because corporate governance is dependent on the functional roles and responsibilities of individuals, this needs to be applied to the execution of formal roles. To some extent, it is possible to do this on your own, but it is more effective if you work with a mentor.

As social beings, the ultimate aim of personal governance must be to enhance interpersonal governance. The purpose of individual roles is to serve the group, and the quality of governance is dependent on the combined contributions of board or committee members.

Thus, Part Three will be devoted to interpersonal governance.

Notes
1. Translated by George Bull (1981) Machiavelli (1515) *The Prince*, Penguin Classics.
2. Bower, S.A. and Bower, G.H. (1981) *Asserting Yourself*, Addison Wesley.
3. Ramsden, P. and Zacharias, J. (1993) *Action Profiling*, Gower.

Part Three

Interpersonal Governance

6   Examples of Interpersonal Governance

▶ ▶ ▶   EXECUTIVE SUMMARY   ◀ ◀ ◀
Part Three of this book explores themes of interpersonal governance, and this chapter consists of a set of case studies of boardroom problems which will provide material for the next chapter's framework for interpersonal governance. These examples are wholly fictitious yet typical, being drawn from my own experience.

While the next chapter sets out to analyze and provide a framework for interpersonal governance, the aim of this chapter is to illustrate some of the infinite variety of ways in which people work, or fail to work, with each other.

Corporations are governed by groups of people – even the most dictatorial of leaders is dependent on others to carry out his or her decisions. Interpersonal governance forms the foundation of corporate governance because the way people relate to each other determines the quality of the decisions they make. What is more, interpersonal relationships between members of separate groups determine the relationships between those groups, and form the foundation of intergroup governance as well.

In the same way that each individual is unique, each group has a unique set of characteristics – combined strengths and weaknesses, habits and quirks. Most boards and executive committees function fairly well despite their ups and downs, but even the best of them seldom function to full capacity. It was suggested in Part Two that more effective personal governance is the first step towards more effective corporate governance, particularly on the human side. But 'no man is an island,' and this chapter illustrates the sort of interpersonal difficulties that can arise in even the most enlightened groups.

While it may be true that the problems described here might have been reduced or avoided if characters had had greater self-awareness, nevertheless, self-governance is not the whole answer. However far one proceeds towards perfecting oneself, there will always be someone capable of 'getting one's back up.' It is a simple fact of human nature, and that is why the importance of seeking others' assistance or opinions was emphasized in Part Two. So, although it was natural to begin with a study of the individual person, the fact is that self-governance and interpersonal governance are in practice closely interwoven.

Although the following case studies by no means exhaust the range of human interaction and groups, it is hoped that they give a flavor and indication of some of the issues. As with the personal case studies in Chapter 3, I am not repeating any true stories but rather fabricating typical examples based on my own considerable experience.

RELATIONSHIPS BETWEEN THE CHAIR AND CEO

Joan Brown had just been appointed chairwoman of an insurance company. She was appointed from outside the company where the CEO, Jacques du Pont, had already held the post for five years. The first six months have not gone smoothly.

The CEO shares his thinking with the chairman and gets little or no response. A week or so later the chairwoman will direct the CEO with little or no explanation, so the CEO has no idea where her directives are coming from. He feels very insecure with this chairwoman. 'I never know what she is thinking. So half the time I guess what she wants, but I really don't know.'

In profiling both people, it became clear that they had very different ways of thinking. The chairwoman, Joan Brown, was quite a private person and needed to go away and think things over on her own before giving an opinion. She also needed to feel good about ideas before supporting them. Jacques Du Pont, however, enjoyed sharing his thinking, and working things out by talking ideas through.

It also became apparent that they could complement each other if they could harmonize their working relationship. Joan was good at seeing how the whole picture fitted together, was able to provide a balanced viewpoint and could help set priorities. Jacques was opportunistic, fast-moving and good at motivating people and getting things done.

After discussing their differences, they worked out an amicable and fruitful way of working together. Jacques would give Joan advance notice that he wanted to discuss an issue with her. He provided the relevant information and agreed a period of time so that she could think about it. Then they would discuss the issue. Joan was able to appreciate Jacques' opportunism, and Jacques felt confident because he felt that Joan would hold him back 'if I get too enthusiastic about the wrong thing.'

SIBLING RIVALRY IN A CORPORATE SETTING

The next story is about a family company where two brothers feuded for many years. As a consequence, the company and the family were divided, and the brothers spent more time sparring with each other than promoting business interests.

The father, who still held the majority of shares and therefore the voting stock, had appointed the eldest son as managing director of the original company. This son had already served the company soundly for a number of years despite having left school with mediocre academic results. The younger son, however, showed precocious talent that won him a place at a prestigious university and went on to the Harvard Business School for advanced business studies.

The company expanded as the eldest brother acquired and integrated small companies into the original one. Meanwhile, the younger brother came back from Harvard and persuaded his father that the company was moving too slowly. The family were impressed by the younger brother's vision, and appointed him CEO of the then growing group.

This deeply upset the older brother who accused his younger brother of usurping his position. He consistently voted against any proposals the younger brother made. This meant that decisions were constantly held up and opportunities lost. The two brothers' wives and families were no longer on speaking terms, and it seemed that none of their children now felt any attachment to the family business.

The mother was extremely distressed by all the bickering, and one Christmas threatened to break off relations with both sons unless they made peace. This shocked them enough to separately take counsel and to work together with a facilitator to see if they could sort out their problems. The first step was to help them see the consequences of their feud. The second was to invite them to test their values and principles, and to explore what could motivate each of them to want reconciliation. When this was found, there was still a great deal of resentment on both sides. Both were able to share their hurt without reacting to each other – although it was clear why the elder brother was upset. He was surprised to learn that the younger brother actually admired his achievements, and had been trying too hard to 'prove himself' in turn. This eventually led to both brothers genuinely being able to understand and forgive each other. They then had to find out whether, and if, they wanted to work together.

In comparing their particular talents and decision-making styles, it was found that although the older brother was not good at communicating his vision and motivating people, he had other strengths. He had a strong sense of the whole company's place in the market, and had built up strong networks and relationships throughout the industry. The younger brother was better at motivating the workforce, and helping directors reach targets. He also had state-of-the-art management techniques, and good ideas for transforming the culture to match current and future needs.

They agreed that, for a period of three years, the older brother would become chairman of the company, and the younger brother would continue to be CEO of the whole group. When disputes flared up in board meetings – 'you can't entirely wipe out bad habits' – they were no longer so serious and were dealt with quickly.

Relationships between the two brothers' families improved, and it looked as though family involvement in the company was likely to continue into the third generation, against all odds.

A CLOSER LOOK AT A GLOBAL FOOD GIANT

After two brief examples of one-to-one interpersonal problems, the next story is that of conflict and resolution within a reasonably good and well-established board. More detail is given to provide the background and to illustrate the relationship between individual profiles and the characteristics of the board as a group.

A well-balanced board like the one described here is likely to be stable and successful for some time. However, some of its weaknesses or divisions may one day put it at a disadvantage, relative to competitors who are more in tune with new markets and product developments.

Nutridelta plc is a food and beverages group that began as a Quaker family company and is now recognized as a major global player and consistent performer in world stock markets. The company benefits from a number of well-established brands plus long-standing relationships with key investors.

There are six non-executive and six executive directors who meet three times a year, plus occasional extraordinary meetings, committee meetings and two-day strategy meetings. The non-executives meet with the chairman before board meetings, and papers are circulated ten days in advance. The non-executives are expected to be available to give advice when needed. As well as the AGM, regular meetings are held with major investors.

Individual contributions – potential and actual

Chairman, Jim Chester

British, 56 years old, chartered accountant and member of the company's founding family. He became finance director, then MD of a large beverage business, and was invited to join Nutridelta as CEO because he stood out as the most competent family member. He is highly competitive, and has achieved distinction as a transatlantic yachtsman.

Particular talents and expertise:
- Understands the advantages of scale, while insisting that component parts work well in detail.
- Able to distill complex data into clear, understandable guidelines, and this motivates both executives and non-executives.
- Insists on high professional standards – information well presented and on schedule.
- Known as an honorable man who is tough but fair. Insists that others are treated with respect.

Leadership style:
- Listens to people's opinions without bias, and checks for understanding, then 'sleeps on' the information before acting.
- Sometimes acts on intuitive hunches.
- Discusses issues off-site with the CEO, selected non-executives and relevant specialists, and presents findings to board as, 'The best solution appears to be as follows, does anyone have any better ideas or problems with this?'
- He then allows a brief debate before summing up and moving on. He hates unnecessarily long meetings.

Main weaknesses:
- Impatience – finishes other people's sentences.
- Unwilling to listen to very different points of view.
- Over-argues in order to win a point.
- Not good at social small talk.

CEO, Pieter de Wolf

Dutch, 49 years old, and a member of the company since graduating in business studies. Worked his way up through several roles, winning several awards for his marketing and advertising campaigns. Also negotiated impressive deals with major customers.

Particular talents and expertise:
- Excellent at acquiring companies and successful mergers.
- Good at integrating groups of people.
- Sets and achieves ambitious targets, especially improving margins.
- Good organizer – grasps details, sees priorities, then pushes others to achieve.

- Highly competitive in the market – understands the competition and sets strategies to outwit it.

Leadership style:
- Highly directive until strategy is determined, but then delegates successfully – 'I like my people walking tall.'
- Inclined to change tack if new information arrives – 'People who can't be flexible don't stay long.'
- Enormous energy and drive.

Main weaknesses:
- Reputation for being cold and calculating.
- Can drive people too hard and precipitate breakdowns.
- Does not always allow things to evolve over time.

Hank Schneider, non-executive director

Well-known banker and specialist in the global implications of IT. Highly informed about the food and beverage industry.

Strengths:
- Asks penetrating questions about the positioning of Nutridelta.
- Good networker – introduces people who might want to sell, merge or create strategic alliances.
- Politically astute – senses people's hidden agendas.

Weaknesses:
- Not sufficiently informed about the group's internal affairs to understand what areas are at risk.
- Tends to take up meeting time with anecdotes.
- Does not often participate in debate.

John Peters, non-executive director

Senior partner in a large consultancy.

Strengths:
- Good at comparisons – especially benchmarking and weighing up merits and demerits of decisions.
- Highly analytical – picks up inconsistency or woolly thinking.
- Intellectually challenging.
- Good strategic thinker.

Main weaknesses:
- Tends to dominate discussion – 'likes the sound of his own voice.'

- Nit-picker – can be aggressive towards executives.
- Not sensitive to people's feelings.

Professor Anna de Silva, non-executive director

Professor of Chemistry at major European university. Is on a European committee studying the implications of genetic engineering.

Strengths:
- Specialist understanding of potential commercial benefits of GM foods.
- Grasp of different ethical viewpoints, and the diverse positions likely to be taken by national governments.
- Sense of humor – able to defuse tensions between disputants.

Weaknesses:
- Difficulty in finalizing decisions – 'seeing both points of view.'
- Can confuse debate with excessive technical detail.
- Lack of business and commercial experience.

Mike Warren, non-executive director

CEO of a major Australian pharmaceutical company and non-executive chairman of a large investment company managing pension funds.

Strengths:
- Development skills – taking highly successful drugs from conception to market.
- Gets straight to the essence of a problem.
- Anticipates hurdles and insists on contingency planning.
- Has made brilliant long-term investments.

Weaknesses:
- Likes to be on prestigious boards for their status value.
- Reads board papers on the lift up to the boardroom.
- Never available to discuss crises when they occur.

Enver Pillay, non-executive director

Qualified as lawyer in his native India and became the CEO of an Asian drinks business that was acquired by Nutridelta. A good financial package plus a place on the board was part of the deal.

Strengths:
- Extensive knowledge of Asian and African markets – many contacts.
- International tax expert.

- Good at finding synergies between companies.

Weaknesses:
- 'What's in it for me?' mentality.
- Quick to take offense.
- Tries to promote his former company rather than work with the group as a whole.

Henk von Solms, non-executive director

Retired CEO of a company that is a major customer.

Strengths:
- Understanding of the retail world – customer focused.
- Knows the importance of branding and 'winning the hearts and minds' of people.

Weaknesses:
- Not always in touch with state-of-the-art thinking.
- Does not always follow the argument – comes up with points already discussed.
- Has strong likes and dislikes among the executive committee.

Combined qualities of non-executives, including chairman

Strengths:
- Experience of different nations and cultures.
- Good relationship with financial markets plus knowledge of different types of investment.
- Shared perspective of the big picture and its component parts.

Weaknesses:
- Arrogance and complacency.
- Not good at 'out of the box' thinking.
- Difficulty with generating alternatives.
- Sometimes fails to anticipate the long-term implications of decisions.
- John Peters and Enver Pillay dislike each other and spar in meetings.

Decision-making style:
- Highly intellectual and analytical.
- Must see the big picture.
- Able to clarify issues and gives sense of direction.
- Action-oriented.
- One member looks for comparisons, one for synergies and one for contingencies.

- High level of debate.

Physical:
- Financially competent, competitive and astute.
- Good sense of practical necessities.

Mental:
- Clear thinking, analysis.
- Able to spot and deal with financial anomalies at speed.
- Intellectually and conceptually extremely bright.

Emotional:
- Stable – not overwhelmed by emotions.
- Considerate and fairly respectful.

Spiritual:
- Implicit values and beliefs.
- Explicit trust that promises are delivered.
- Competitive, critical and optimistic state of being.

Vignette

At the time of writing, the chairman is thinking of retiring in two years, but is not sure whether the CEO should be his successor. 'He is good at getting results, but I'm not convinced he has the diplomacy to be chairman.'

Although the non-executives work well together, the chairman is an advocate of continuous improvement. 'I'm not sure if I'm getting the best from them or whether we are doing as well as we could.' He is frustrated at the amount of time John and Enver waste 'just to score points off each other,' and he feels that the CEO 'keeps a tight ship and is not happy if an executive director comes up with alternative proposals.'

In addition to the normal business of the board, three issues have been taking a lot of its attention. The first was a closely averted scandal stemming from an unfortunate ethical branding project. In face of rising public resentment at the power of global food chains, Anna de Silva and Henk von Solms proposed a collaborative venture with a charitable anti-drug organization to offer grants to coca (the raw material used for manufacturing cocaine) farmers to switch to coffee growing. After initial resistance from the board, it was seen that this could tie in well with their recent acquisition of Enver Pillay's company and its international chain of tea and coffeehouses. The One World Coffee Company became a highly successful global brand with its youthful appeal and strong ethical associations. It also delivered higher profits, as the resulting growth in coffee production helped lower the cost of raw materials. Unfortunately, this very success contributed to a slump in coffee prices. This not only tarnished the ethical aim of 'helping Third World farmers' but also led to a reversal of the decline in coca production as impoverished farmers were now planting coca to tide them over the recession, knowing that they could then claim their grant for removing coca bushes – maybe for the second time. While the majority of the board agreed it was time to rethink the strategy, Enver Pillay and Mike Warren lobbied so insistently for its retention that John Peters let it be known that he suspected Enver of being in the pay of the coca barons. This led to considerable hard feeling which was only resolved when discrete inquiries absolved Enver from any such involvement, and he agreed to support a re-launch of the One World Coffee brand as a 'healthy family alternative' in bars serving alcohol.

The second, less divisive, issue is succession for Hank Schneider – due to retire at the next AGM. The nominations committee wants agreement from the board about criteria for selection.

The third issue is another ethical 'hot potato' – the question of genetically modified foods. The board is divided on this, having burned its fingers by investing heavily in producing genetically modified foods and then having to slow down in the face of public pressure. Although members were already well informed via Anna, they have agreed to ask specialists to update them on national policies and practices across the globe. The chairman said, 'I feel a little stuck about what to do next. The executives want serious investment in GM foods now rather than miss the window of opportunity. I think this could be premature.'

The executive committee

In addition to the CEO, the committee includes the following directors: finance, production, logistics, marketing and human resources. Although all are board directors, they provide a united front and loyalty towards the CEO.

Apart from answering questions related to their respective functions, they tend not to contribute a great deal to the board meetings. Typically they give presentations and recommendations which are then challenged and discussed by the non-executives.

Two non-executives act as mentors to Sally Brown, the logistics director, and Dick Bradley, the finance director.

By and large, relationships are constructive, and non-executive directors are satisfied with the timeliness, quantity and quality of information given. There are criticisms that some presentations are too slick, and make it difficult for non-executives to add value.

Main strengths of the executive committee:
• Homework is always well done.
• High intellectual caliber.
• Each director is competent, professional and delivers on target.
• Systems and processes work effectively worldwide.
• Marketing and sales are aggressive, fast-moving and continuously growing.
• Mergers and acquisitions are successfully integrated into the company.

Weaknesses of the executive committee:
• Only the CEO and the logistics director have a shared perspective of the company as a whole. Directors tend to defend their own functions rather than act for the group.
• Lack of lateral 'out of the box' thinking.
• No long-term vision.
• Lack of contingencies for possible crises.
• Low motivation to stand back and reflect on how they could develop as a team.

Strengths of the board as a whole:
• Highly professional, intellectually rigorous and analytical.
• Well-argued, challenging and constructive debates.
• Highly respected by the financial markets.

Weaknesses of the board as a whole:
• Arrogance and complacency.
• Lack of 'out of the box' thinking.
• Poor capacity to respond flexibly to unexpected threats or new competition.

Comment

This board is reasonably healthy and successful. Shareholders are pleased with the overall results. It is probably working at about 60 per cent of its capacity. If more work had been done on agreeing on shared values, it would have been easier and quicker to deal with the coca problem. A great deal of time, energy and goodwill could have been saved if the chairman had been more skillful in stopping the conflict between John and Enver. The greatest danger, however, is the sustained overall success of the company. There is an atmosphere of complacency that could put the shareholders' money at risk in the long term. The board may overlook new markets and new forms of competition that could overtake the company when they least expect it.

One problem this company did not suffer was disagreement between nationalities. Jim Chester claimed that he and Pieter were 'both from old colonial countries, with a tradition of keeping peace within our empires.' Be that as it may, the subject of our next case study was not so blessed.

GLOBAL NON-ACCORD AT BANK OLYMPIA

Bank Olympia was the product of an extended merger between French, Dutch, German, Swiss and American banks and insurance companies. It was failing because board members could not get on with each other. Also, the board was so large – 35 people – that it was difficult for the Swiss chairman to handle it. Each group had an equal share so it was difficult for one group to dominate another.

There had been a honeymoon period for two years, when everyone co-operated to achieve the economies of scale that had inspired the mergers and to improve margins by cutting costs. Financial results were extremely healthy. The trouble began after other banks went through the same merger process, and Bank Olympia's competitive advantage was narrowing.

The Swiss chairman spent long periods of time with the company secretary constructing an agenda to discuss strategy, and he expected it to be rigidly adhered to. The Americans, however, wanted to be involved in designing the agenda, and wanted more time for other issues they wanted to raise. Every meeting began with a complaint by the Americans who asked how they could cooperate if their items were not given enough time. Meanwhile, the Dutch did not like the Germans and accused them of Gestapo-like activities when visiting Dutch banks. In addition to this, two Frenchmen were having difficulties:

Pierre complained that Dominique was 'an arrogant sneering école supérieur idiot who thinks he can mince his way across life with the great and the good.' Dominique disapproved of Pierre and accused him of racist and xenophobic tendencies that would undermine the group's ability to take advantage of synergy.

There was talk of de-merging. At the time of writing, the chairman is trying to work out how to deal with the various difficulties. 'It is quite obvious to me. Why can't they see it my way?'

Comment

Many corporate governors are accustomed to working in an international environment. However, problems caused by clashes of national temperaments are well known. (1) What would have helped this group would have been to have checked out with each national group how they preferred to work and what their expectations were. It could also have helped the chairman to bring in advisers who are specialists in this field.

REBUILDING THE EXECUTIVE AT GANDOR AND ORPWOOD

So far the stories have related to main or supervisory boards. This story and the next concern an executive committee. Some of the problems are common to both groups.

Gandor and Orpwood had a large business machinery maintenance division employing some thirty thousand engineers. They were not doing well when a new chairman, Ben Smith, came in and decided that the company had attitudes and practices that 'go back to the nineteenth century.' He felt that Jack, the CEO, behaved like a dictator, and did not give his senior managers scope for initiative. 'I called him in and told him that he had six months to change his style, and to come up with some innovative ideas to wake the business up.'

After much resistance and soul searching, Jack decided that he was going to grasp the challenge. He took himself off to a prestigious business school for a course in team building and hired a consultant to help him transform the organization. He had profiles made of his key senior managers to see if he had missed any good potential. Most importantly, he told his people that he was changing his style and that he expected the same from his senior managers.

He had a vision of creating a flat structure with self-managing teams, and became enthusiastic about self-responsibility and accountability, deciding to 'give more responsibility to my guys.' At the first meeting he sat in silence without saying anything. Nothing much happened at the meeting, other than one person saying, 'If you tell us what you want, we will do it.' He realized that if he wanted the rules to be changed, it would help if he spelled out his expectations and how they related to the needs of the business 'for a rebirth.'

'The first thing is to come up with creative ideas.' He took a group of his top people on a ten-day trip around the world to look at best practices. People throughout the organization were invited to contribute ideas. He then had specialists take all the senior managers through a series of highly creative thinking exercises and techniques.

There was enormous resistance at first: 'I don't want to make a fool of myself.' 'You don't seriously expect us to behave like nursery school children and draw pictures?' Some of them stood around with arms folded and lips pursed. However, by the end of two months, 40 worthwhile ideas had been generated. These were distilled down to five, which could be tested by a group of major customers and suppliers. Better relationships were immediately forged, because they enjoyed being consulted and found the process enjoyable.

What resulted was a decision to change direction, and for the engineers to act more as consultants for their customers' ongoing business machinery needs. They helped with ordering machinery, maintaining it and then periodically upgrading it.

Some people still felt that the creative work was a waste of time. However, it did set the atmosphere and state of being for the company to embrace change enthusiastically.

Intensive training followed, beginning with the top 20 managers. 'There will be no different titles. You will all be directors and, as a team, will be accountable for the results.' They learned how to operate effectively as a large group as well as working in sub-groups. Each director developed effective leadership and team-building skills, which cascaded down through the organization. They had clear policies and principles that were practiced consistently throughout the organization. Every director was capable of presenting the same message to the board.

The CEO was delighted when, two years later, he was promoted elsewhere. 'I can leave this company in many excellent hands,' he said.

Comment

This story started with a chairman knowing clearly what he wanted. The CEO was willing to change, and threw himself enthusiastically into transforming his company. He also motivated old-fashioned conservative engineers to not only change their behavior, but their approach to business. They became entrepreneurs rather than maintenance men. As a result, the company did extremely well, and is still highly profitable.

Not all stories have a happy ending. Not all groups work well. The next story is about a relatively new company that has achieved meteoric success in the stock market, but is dealing with a relatively unproven product. The story is pure fiction. However, the scene described is not uncommon.

THE STORY OF THE EXECUTIVE COMMITTEE OF YMT

YMT was founded by Tony Petronelli, an ambitious CEO with an ambitious vision. He thought that he was close to discovering the ultimate cure for migraine. 'This will reduce the amount of human suffering, and save industry a fortune in days lost at work because of migraine.' The idea came from his previous company – a pharmaceutical giant that had cut back its research and development team. Some members of the team had been extremely excited by the discovery of an extract from a lime tree that seemed to have miraculous effects on rats suffering from conditions similar to human migraine. Because the company was winding down the department, they did not pursue this discovery and Tony was sure he saw a golden opportunity.

He managed to get a large medical foundation to raise money to take on the team and continue with the research. This went well enough for the company to go public. The chairman, Peter, who was well known and had been CEO of a large company, was inspired by Tony. 'I have always wanted to put my name to a brilliant discovery like this.'

They recruited a young finance director, Marco Viti, who was brilliant in the City. Together with the enthusiasm of the CEO and the support of the chairman, they raised more money than they ever imagined and looked all set to make a killing. Marco was extremely ambitious, and wanted to be in the list of the world's top two hundred wealthy people by the age of 35.

The trouble started when Marco put the risk analysis on the agenda. Tony put the item at the end of the agenda, and no matter how many times Marco brought it up, it was never discussed. He tried to talk to Tony in private, but Tony always changed the subject. At the same time, David Lee, the research director, was becoming a little anxious about the pressure to produce results within a short time span. Again, Tony fobbed him off. He and Marco got together and decided to bring the subject up together at the next committee meeting.

MARCO: 'Before we discuss other items on the agenda, could David and I have twenty minutes to discuss our concerns?'

TONY: 'We have more important things to discuss right now. Can we get on with ... .'

MARCO (loses his temper): 'There will be nothing to discuss if we don't talk about risk.'

TONY: 'I know what I am doing. Why can't you trust me?'

MARCO: 'That's not the point. We need to talk about risk.'

DAVID: 'I agree, what will happen if ... .'

TONY: 'What's that got to do with you? I want you to know that I feel very upset. I was expecting more loyalty from my team.'

Tony walks off in anger, slamming the door. The team sits in amazed silence.

This scene was repeated almost verbatim a number of times.

A few months later, the trials were not going too well. Marco was so angry that he held a meeting with selected investors and members of the press without telling the chairman and CEO. Then the research director publicly resigned. Share prices fell sharply. Tony and Peter resigned, and Marco was not accepted back into the City as he had gained a reputation 'for being too greedy, even for the City.'

Comment

If Tony had respected the concerns of his directors early enough and had listened to them, the problems that arose would have been alleviated. The fact is that he had thought about the risks and had contingencies, but had not communicated this to his directors. The problems were exacerbated because of emotional immaturity, on the part of both the CEO and the finance director. The research director was affected by the general atmosphere of discontent, and followed the destructive path of the finance director.

SUMMARY AND CONCLUSIONS

In this chapter, a number of stories were invented to illustrate some of the ways groups operate and what the human consequences are. Each story is different. However, there are also certain components of group behavior that are almost universal. These relate to the process of collective decision-making. These components are described in Chapter 7, A Model for Interpersonal Governance, which will be followed in Chapter 8 by Guidelines for Interpersonal Governance. Chapter 9 looks at a topic that has concerned corporate governors for many years – The Shadow Side of Corporate Governance. Chapter 10 looks at the implications of both personal and interpersonal governance as they are related to the particular functions of the main or supervisory board, the executive committee and the individual roles within them.

Note

1. Marx, E (1999) *Breaking Through Culture Shock*, Cambridge University Press.

7   A Model for Interpersonal Governance

Business prosperity cannot be commanded. People, team-
work, leadership, enterprise, experience and skills are what
really produce prosperity. There is no single formula to weld
these together, and it is dangerous to encourage the belief
that rules and regulations about structure will deliver
success. (1)

▶ ▶ ▶   EXECUTIVE SUMMARY   ◀ ◀ ◀

In Chapter 6, examples were given of different kinds of interpersonal governance.

This chapter outlines certain dynamics that arise within any group engaged in corporate governance, whether supervisory or main boards, or executive committees. This model is based on observing successful groups and extracting common components with a focus on collective decision-making, as this is considered one of the main functions of governing groups.

The subject is addressed under three headings: content, roles and process. Content includes the need for prior information and understanding of a topic to make intelligent decisions about it. It also underlines the fact that any major decision should always take into account the human face of corporate governance, and that requires people to pay attention to the way they themselves are treated and the way they treat others. The roles discussed are those of the leader (directing and facilitating), the participator and the influencer. Under process, six stages are considered: self-understanding; awareness of the group; design and planning of the decision process; enjoying the actual process; ensuring authentic commitment; and follow-through to results.

This chapter intends to outline a model for interpersonal governance that gives the reader a systematic way of thinking about how decisions can be effectively made (Figure 7.1). It is an attempt to simplify what is actually a highly complex and dynamic process. The model is based on observations about how

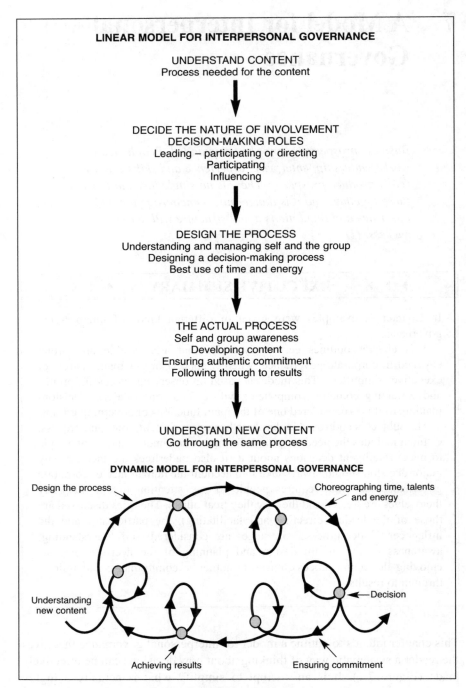

Fig. 7.1   Model for effective interpersonal governance

successful corporate governors operate. It also takes into account advice given by outstanding board members in terms of what they have found helpful. All of this has been gathered together in a framework that the reader will hopefully find of use. The focus is less on theory and more on what people find is of practical use. This will be further developed in the next chapter on Guidelines for Interpersonal Governance. At one level, the model suggests a linear approach to interpersonal governance. On another, all these aspects are seen happening simultaneously whenever a group of people is thinking about a topic or a decision.

The topics that are explored in this chapter are therefore expressed both in a linear fashion and as a dynamic and evolving process.

UNDERSTANDING CONTENT

One of the great complaints of board members is that they do not have sufficient time to absorb important information needed for making decisions. This information is often produced in a format that is either too detailed or too sparse. One non-executive director commented, 'What I want to know is what I need to know to make a decision. This means that it is useful to me to know why I am reading this stuff and what the key issues are.'

It helps when people have done their homework first. Understanding content at a deeper level comes through discussion and the value that directors add to the original information. It is this shared understanding of the information and the implications that form the basis of successful decisions.

The reasons for the decision

It may sound silly, but I have sat in on meetings when members have not been sure why they are making a decision. It is so easy to get caught up in the corporate timetable and to go mechanically through habitual items on the agenda. The non-executive who asks 'why?' usually helps the board to refocus on what is important. At the risk of sounding trite, leaders who remind people of the reasons and timeliness of a decision save a great deal of time and confusion. This is particularly useful if there has been a significant delay between meetings.

Quality and timeliness of information

Information cannot be adequately digested and absorbed if it is handed out at short notice. Too much, or too little, information can also be used by the CEO to prevent directors from challenging their proposals. This means that, although the CEO may get what they want, they are hindering directors from carrying out their roles. Nor, with this procedure, can issues be intelligently discussed and better solutions suggested. A good example of this is given in Chapter 6 when the CEO of YMT refused to discuss risk with his directors.

Most directors prefer a short summary of the relevant information with guidelines on how to read it, the key issues and what is expected of them. Some prefer to have a number of options spelled out with their respective advantages and disadvantages. Directors generally feel more secure if they are informed of the difficulties as well as the good news well ahead of time. Neither investors nor directors like unexpected surprises, especially if they are unpleasant.

Each person tends to absorb and deal with information in different ways. It is also useful to know those ways in which board members do not like to have information presented – not everyone is comfortable with the written word, and many prefer to be briefed verbally. Where the trouble is taken to find out individual needs in this respect, it is usually not too difficult to provide adequately for each board member, even with large boards or committees.

Development of content

In the best groups, exploration of content is a creative process in which it is ingested, discussed and debated. People share additional information, clarify issues, point out different perspectives and suggest alternatives. By looking at issues from different angles, they challenge thinking and identify additional work that needs to be done before a decision is reached.

The original content becomes a starting point for a journey that leads to a successful, or unsuccessful, resolution. This is one of the main skills of a good facilitator – to elicit the best input from directors, pull together arguments and suggestions and come up with a summary that accurately represents the individual and combined points of view. In this sense therefore, content can be constantly changed and improved until a final decision is made.

The content must include the human side

In the first two chapters, the question was asked, 'What is being governed?' It was suggested that the 'human side' must embrace everything that people do, but all too often it is only the financial aspects that are considered. If groups take a little extra time to consider the fuller human implications of their decisions, they could make a positive difference to the lives of the people they affect. Those who fail to consider the human implications are arguably too socially irresponsible to fulfill their roles.

DECIDE THE NATURE OF INVOLVEMENT: DECISION-MAKING ROLES

People involved in governance often serve on a number of boards or committees and may play different roles on each. Because of this, I have looked at the different characteristics of each role and what is required. Three main roles are considered:

- The leader.
- The participator.
- The influencer.

The leader

Leadership style depends on three major factors. The first is the values, beliefs and expectations of the leader and his or her group. This is mainly a political and ideological issue, even though the way in which people govern is taken for granted. The predominant ideology worldwide appears as a blend of market economy, competition, and democratic values. The second factor is the personality of the leader and the way that they are naturally predisposed to lead – many show a need to dominate, control and display power. In practice, leadership styles embrace the full spectrum of possibilities, from pure industrial democracy to downright authoritarianism, with the majority somewhere in between. The third factor is the type of leadership or participation that an actual task demands.

When I started working in industry 20 years ago, I believed strongly in promoting industrial democracy. I have since learned just how difficult it is in practice. It demands high levels of personal responsibility, accountability, understanding and an ability both to challenge and to offer ideas. A democratic leader's role, at its highest level, is to coordinate and represent the views of the majority. It is more the role of a facilitator than a dictator, and is often what an excellent chairman does naturally on boards where there is consensus.

I also learned that, although absolute tyranny is unacceptable, there are times when strong direction is appropriate – for example, during times of crisis or when urgent opportunities need to be seized. Most groups do not move quickly enough to take opportunistic decisions. Although I would prefer an individual to be mandated by the group to take responsibility at such times, this is not realistic. I therefore accept that, although the democratic process is ultimately preferable, the majority of leaders need to adjust their roles to suit the situation: sometimes emphasizing their role as facilitator (leaning more towards democracy) and at other times their role as director (leaning more towards authority and control). The chairman of Nutridelta had a natural inclination towards facilitating and arriving at a democratic consensus. He could, however, have been more prescriptive in not allowing John and Enver to fight with each other at board meetings.

Where the leader plays both these roles, it can be difficult to change gear. The important thing is to be clear with the group about the role, and what is expected from them. Both types need to understand and to be able to handle the shadow side of interpersonal governance, and to be able to deal with normal interpersonal difficulties – personality clashes as seen in Chapter 6, people who are not performing, people who are dishonest and many others. This shadow side is described in greater detail in Chapter 9.

The leader as facilitator

Good facilitators are those who enable others to take full responsibility for their thoughts and actions. They provide an environment in which individuals and the group as a whole can co-create decisions and activities. Good facilitators know exactly when to invite any individual to contribute, because they understand each person and their individual skills. They welcome members' initiatives for improving the process and, on some occasions, to assume temporary leadership of it. A good facilitator knows when and how to guide people towards consensus. They usually encourage a learning environment for the group. They influence rather than prescribe; coordinate rather than control. Often, an effective facilitator is invisible, and the group believes it has done all the work itself.

Good facilitation can deliver the advantages of self-governing, proactive people who individually and collectively take full responsibility and accountability for their decisions and actions. Individuals have the freedom to contribute within a group that has become powerful and cohesive. A shared sense of purpose and capability can move mountains.

The disadvantage of facilitation is that it can allow the less proactive members to be carried by the more proactive ones. It can also take a great deal of time to reach decisions if the facilitator, or members of the group, are not developed enough to take responsibility for initiating appropriate contributions. Good facilitation requires cooperation, intelligence and advanced social skills that not all groups possess. It is a learning process that takes time to develop.

Another problem may be that a powerful facilitator can manipulate people into arriving at conclusions that they have already reached. This leaves people feeling uneasy and cheated.

The leader as director

Good directors are able to take responsibility for the development of content as well as process. They often formulate solutions and then sound them out with the group before acting on them. They are decisive, and confident that they will succeed. Good directors are usually in control of what is happening even when they delegate. A good director can inspire and challenge people to produce work beyond anything they thought they were capable of. They provide clear direction and steer people along the way. They are often highly astute, intelligent and respected and command loyalty and following. A good director can explain why a decision should or should not be made.

The advantage of a good director is that clear decisions can be made and acted upon with speed. People will know their place and will know what to do. If they toe the line, they are likely to be looked after. The art of 'followship' is less effort than self-governance.

The disadvantage of directing is that 'followship' seldom breeds people willing and able to direct or facilitate. When such a director leaves, there can be a vacuum which leaves a board or an organization at risk. Another disadvantage is that there are seldom enough strong people to challenge the thinking of the director or mitigate their weaknesses and/or blind spots. It is easier for them to make limited decisions or errors of judgment.

Another danger of directing is that it can lead to excessive power and tyranny that undermine people's dignity and confidence. People can be motivated by fear and defensiveness, covering up and hiding mistakes that in the long run can put a corporation at risk.

The role of the participator

In Chapter 6, examples of interpersonal governance were offered to illustrate some of the things that go on in boards and committees. By now, it should be obvious that the role of the board or committee member who participates rather than leads is vital. Participation is an art form in its own right. It is not just about attending meetings and participating when this seems a good idea. The quality of participation often makes the difference between a good and a poor decision.

The most common complaint from a leader is that their directors are not pulling their weight enough. The task of the participator is first to understand the nature of the content and what is needed from them. This probably entails taking the initiative to ensure that the right information is offered well in advance. Second, it is up to the participator to understand their key strengths and to decide ways in which they can best contribute to a particular decision.

They then need to determine how and when to contribute and when not to contribute. This involves understanding the needs of the leader and how the group as a whole, as well as individual directors, are likely to operate. It is a matter of knowing what to contribute and when to intervene, as well as finding the most appropriate way and time.

Any participator can contribute in a number of ways, depending on what is needed and what they can offer to the process. This can include sharing their physical, mental, emotional or spiritual intelligence. They may be requested to give information, offer alternatives, clarify issues, discover priorities, compare options or examine the implications of decisions. They may also be required to ask difficult and penetrating questions, challenge thinking and help others to express their feelings, or diffuse tension through humor. Often one director acts as moral guardian for the rest of the group. The list is endless.

There are also many ways in which participants can undermine the process. For example, some arrive late, without having done their homework. They can try to dominate meetings, try to compete with other directors, play games of one-upmanship, tell anecdotes, fall asleep, complain about each other behind their backs or argue with the leader. One of the major failures in terms of the quality of participation is the inability to listen in order to understand. This causes misunderstandings and people to talk at cross-purposes.

While some directly participate, others try to influence the thinking and behavior of key people.

The role of the influencer

Everyone knows that a great deal of governance takes place off-site in restaurants, corridors, golf clubs, planes and any other place where people meet. There is a great deal of canvassing and attempts to influence thinking. Networking can sometimes be effective in terms of swaying opinion. Not only do directors try to influence each other, but people further down the organization can carry a great deal of weight. This is particularly true where boards are extremely large. In this case, companies are often run *de facto* by senior managers who are sometimes not even on the executive committee. On a more sinister note, there is the dark side of influence – bribery and corruption.

Table 7.1 gives some indication about differences between outstanding, strong, mediocre and weak leaders, participators and influencers. A distinction is made between leaders who act as facilitators or directors.

| | LEAD | | ADD VALUE | |
	DIRECTING	FACILITATING	PARTICIPATING	INFLUENCING
OUTSTANDING	• Active listener. • Mandated to be directive. • Inspirational. • Can both direct and facilitate when needed. • Uses input of others. • Speed, decisive. • Makes things happen.	• Democratic approach. • Active listener. • Creates appropriate environment for achieving good outcomes. • Enables others to give of their best. • Can direct if necessary. • Team builder.	• Active listener. • Well-informed. • Asks excellent questions. • Appropriate interventions. • High-level debating abilities. • Guardian of certain parts of the process.	• Active listener. • Receives good information. • Highly trusted. • People seek them out. • Reciprocity of ideas, information. • Effective in helping people work out their ideas.
GOOD	• Trusted and respected. • Able to control both content and process. • Uses others to test ideas and add value. • Prepared to change their minds.	• Aware of and manages own and others' strengths and weaknesses. • Creates high-level, trust and learning environment for input and debate. • Highly skilled and trained. • Knows how to co-ordinate, summarize and complete.	• Aware of own strengths and when best to contribute. • Collaborative in approach. • High level of debating ability. • Knows when to keep quiet. • Well-informed.	• Politically astute. Knows where to go. Effective. • Creates effective network of people. • Gets things done behind the scenes. • Shares information.
MEDIOCRE	• Inconsistent behavior. • Vague directions. • Commands just enough trust and respect.	• Not always clear about purpose or desired outcomes. • Mechanical or repetitive. • Sometimes aware of the group. • Inadequate skills.	• Expecting to be told what to do much of the time. • Occasionally adds value on a random basis. • Sometimes operates on too low a level.	• Lacks a sense of power. • Makes some attempts to influence. • Can give up easily. • Can lack skills/information or know who to influence.
POOR	• Tyranny – creating fear, distrust and defensiveness. • Poor judgments. No one to challenge them or add value. • Picking up and dropping favorites or • Indecisive and surly.	• Little or no facilitation or Interference all the time. • Allowing discussions to go around in circles, never arriving at decisions. • Not able to recognize or deal with the shadow side, conflict or sabotage.	• Fear or apathy or anger. • Almost no participation, or inappropriate interventions. • Sabotage – conscious or unconscious. • Disruptive behavior. • One or two members dominate and take up too much group time.	• Fear or need for power. • Apathy or malevolence. • Almost no influence. • Playing 'politics' – causing divisions and rumors. • Undermining initiatives or sense of impotence (just gives up).

Table 7.1 Collective decision-making. Levels and types of involvement

THE INTERPERSONAL PROCESS

Good leaders tend to inspire the groups they lead. Many of the best decisions stem from the way the issues are tackled. Where a group of people works well together, they will usually say something on the lines of 'the best boards are those that are the most fun. We get down to the essence of an issue and then we thrash it out, no holds barred.'

The atmosphere of a motivated group of people who enjoy the process is very different from one that is apathetic or confused. However, it is also possible to have a motivated group which does not make good decisions because it does not have the skills or the rigor to deal sufficiently well with the content. All these factors are necessary to obtain the best results.

The term 'process' is used to describe the ways in which decisions are reached. Two aspects are considered. The first has to do with the **interpersonal dynamics** of the group itself, the unique ways in which the group behaves, how people relate to each other and how values and beliefs are acted out in the group.

The second aspect concerns the actual **decision process**, that is, how a group makes decisions. This is influenced by the nature of the decision as well as the style of the person leading the group. A more directive, decisive and logical person is likely to lead discussions in that manner. A more facilitative, easy-going, creative person is more likely to leave things open-ended until the last moment. The values and beliefs of the leader are likely to have a significant effect on the ways in which group members are seen and treated, and indeed how that person treats him or herself. If a leader does not fundamentally know their own value, then how can they truly evaluate others?

It has been found that the quality of the human face of governance grows with the quality of relationships within the governing group. As was indicated in the chapter on personal governance, this starts with the relationship each person has with themselves as well as their beliefs about human nature. Groups whose members treat each other with respect, acknowledgment, trust, honesty and fairness are more likely to care about the human implications of their decisions. They are also the most likely to have clear policies concerning social responsibility and other ethical considerations.

True, most people are clear about how they want to be treated and how they would like to treat others. However, when things get difficult, it is not always easy to behave impeccably. Being mindful of one's effect on others is easy to say but not always easy to do.

Aspects of the decision process

There are a number of elements that contribute to successful decision-making. They do not all need to be present at the same time, nor need they occur in the order given below. In any one cycle of decision-making, any or none of these could be manifest. An improvement in any one of these is likely to have a positive impact on the quality of decision-making. The aspects are as follows:

- Self-understanding.
- Awareness of the group.
- Design the process.
- Enjoy the process.
- Ensure authentic commitment.
- Follow through to results.

Self-understanding

Part Two of this book focuses on self-awareness and self-development. The more someone understands their unique talents and expertise, their decision-making style, their strengths and weaknesses, the more able they will be to contribute appropriately to the group. I have discovered that if each board member shares with others the value they would like to add to the group, and in what areas, the more other members are able to give of their best.

It also helps if they share their preferred ways of working, their needs in terms of the amount and format of information they would like as well as those things they do not like. If someone hates being given things at the last moment, then it is prudent to give them information ahead of time. However, if another person responds better to working to deadlines under pressure, they may not mind being given information at the last minute.

It also helps to be aware of the way in which one interacts with others. Some people like to share their thinking and feelings. Others need to go away and think, and then come back with their views. Yet others may switch from sharing their thoughts and feelings to 'going private.' Each person is good at some form of relationship and has difficulty with others. One chairman is blunt and says what he thinks and feels. This can be useful because his board members know exactly where they stand with him. On the other hand, as he says, 'I do not exactly get the prize for tact or a light touch.'

Most people have at least one fatal flaw – their Achilles' heel – that can cause them or others difficulties. In Chapter 3, Jane did not like confrontation, and the result was that an underperforming CEO nearly put the company at risk. Jane was able, through hard work and determination, to overcome her difficulty.

However, some fatal flaws, more deeply entrenched or hard-wired, are more difficult to change. A chairman of a global company said, 'no one is perfect. Perhaps the main task of the chairman is to support the CEO to not get into trouble. An important task of my non-executives is not to allow me to indulge in my excesses.' A CEO commented 'as a team we compensate for each other's weaknesses. It is lucky that we do not all have the same weaknesses.'

One of the most important aspects of self-understanding is awareness of how each person manages their time and energy. A common complaint is that there is not enough time for people to reflect on what is really important. 'I am too busy doing my own job, traveling and wasting time on irrelevant meetings and committees.' The fact is that many directors are action-oriented, and not naturally motivated to take time to reflect, even though they know they should. The effective management of limited time is, however, vital for effective governance.

Awareness of the group

In some groups, people work together over many years and get to know each other fairly well. However, it is also possible to attend regular meetings with the same people and never gain more than a superficial knowledge of them. People tend to slip into patterns of behavior over time that may not express their true potential. The more informed each member is about the different capacities of each person and the dynamics of the group, the better able they are to know how to relate to other directors and the group as a whole.

When I ask board members to comment on their fellow directors, many already show an astute understanding of them. However, this is often instinctive, and not often leveraged in a conscious way. There are four levels of understanding that any board member can develop:

- Understanding the strengths and weaknesses of each director, including their talents, expertise and decision-making style.
- Understanding the style and preferred ways of working of the chairman and CEO. Also, understanding their own strengths, weaknesses and how best to relate to them.
- Understanding the intrinsic qualities of the group as a whole, what I call the group's 'Inner Brand.' This includes the way the group combines individual contributions, and the predominant values and beliefs that structure its behavior. It also includes an understanding of the actual and potential capacity, knowledge, expertise and skills of the group to do its job. In my experience, even highly competent groups only perform at about 60 per cent of their capacity.

- Understanding the dynamics of the group. This has to do with the way any group arrives at decisions, the quality of meetings and how the group manages its time. It is the actual process and methodology by which decisions are made. It also relates to the way in which the group spends or wastes its time.

DESIGN THE PROCESS

To some extent, everyone taking part in a governance group will prepare for a meeting and think about what they want to say and do, whether they are leading the group or participating in it. Few people are so talented or experienced they can go into a group and improvise and produce stunning results. Although this does happen, groups where people pre-plan the process do usually produce better results. Using this approach, the way meetings or projects are run can be fine-tuned both to address the challenge of the content and to utilize more effectively the strengths of the participants. Figure 7.2 shows an example of the relative strengths and weaknesses of a group's decision-making.

An understanding of the content and its purpose will not only provide focus, it will also allow the selection of appropriate methods. Most leaders run their meetings in the same way. However, it is possible to develop a repertoire of different ways of making decisions that may be more appropriate to the matter in hand. For example, some decisions require careful analysis and detailed information. Others may require lateral thinking to generate alternatives. A situation requiring people to formulate ethical policies demands an approach that taps into emotional and spiritual intelligence and enables people to say what they feel.

Also, different types of groups may respond better to different types of meetings, or may need different methods to get them out of bad habits. One example is an executive group which had selected themselves because they were interested in working in a company alongside people with similar values. They were excellent when it came to creating long-term visions or defining a shared set of values, but they could spend hours simply debating the meaning of everything. The CEO had to find ways of enabling them to reach timely decisions so that the rest of the organization could get on with its activities.

Another group was so caught up with the financial aspects that it neglected to check whether the people in the organization were capable of reaching the targets it had set. Not only was management incompetent, but the low morale in the organization meant that there was little motivation to achieve goals. The chairman had to work together with the CEO, first of all to improve motivation,

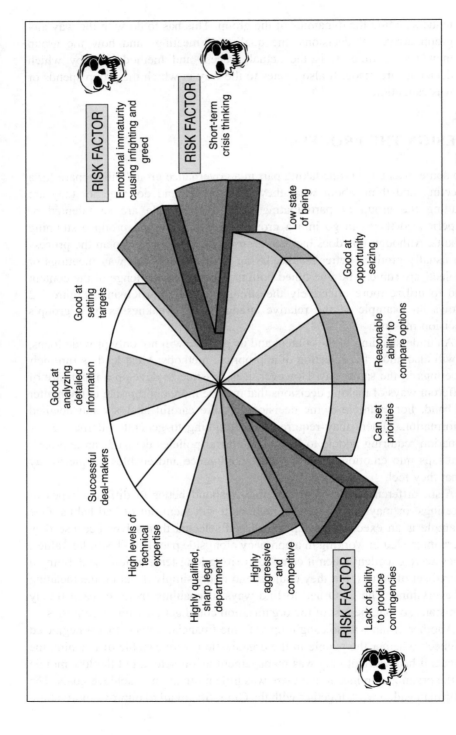

Fig. 7.2 Example of an opportunistic retail board. Relative strengths and weaknesses of a group's decision-making

and then competence. He had to include items in the board agenda that addressed these 'people issues.' The story of Gandor and Orpwood is a good example of this where the CEO radically changed his own style to enable the organization to transform.

The size of the group also dictates how decisions are made. In Chapter 6, in the first example, the chairman mainly worked with a small group of non-executives. Discussions could be intimate, and each person could truly contribute as much as they wanted. Meetings could be informal, and board members got to know each other fairly well. In the executive group of 20, where a flat structure was used, different dynamics had to be employed. The CEO had to trade off the time he spent with individual directors with the time spent with the group as a whole. To some extent, the same consideration applies to the large board of 35 people – although one group deals with executive issues, and the other with global governance issues. In large groups, people tend to split into sub-groups. It is also possible for chairmen and CEOs to develop effective skills for enabling large groups of individuals to make highly participative decisions if this is required. This can be applied to addressing large groups of people, for example at the AGM.

The main purpose of planning a process in advance is to find the optimum ways in which the group could combine its talents, knowledge and experience to reach the best solutions. A major consideration is to do this with the minimum expenditure of time and energy. This requires the leader to have the talents of an orchestral conductor so that the performance is greater than the sum of its parts. The leaders also need to be able to choreograph activities within clearly defined time limits. Effective planning of how to use time is a great asset.

THE ACTUAL PROCESS

Enjoy the process

The majority of investors, shareholders, board members and executive directors would not be there unless they enjoyed the process in some measure. However, there is scope for the process to be significantly more enjoyable and effective than it is in many of the groups that I have observed.

The reason is that the scope of what can be done in boards and executive committees has barely been tapped. Most people involved in governance lead some groups, participate in others, and on others make the biggest impact by knowing how to influence key people. An individual can therefore have many roles at any one time. When one learns to enjoy or 'play' these roles, rather than simply 'fall into' them, then they become roles of one's own choice. This, in turn, leads to a stronger and more authentic sense of commitment.

Ensure authentic commitment

So far, I have discussed the importance of understanding content and the different aspects of the decision process. However, one of the most important and sometimes overlooked ingredients of effective governance is knowing whether people are truly committed to a decision or merely paying it lip service.

Where money is concerned, particularly when there are high stakes and ownership, there is a likelihood of high levels of commitment. The difficulties more often arise when it comes to policies that affect the way people are treated, both inside organizations and in the community. Sometimes people will allow themselves to be treated with disrespect if it means making money. However, they seldom stand back to consider the real price of selling their souls.

Unless there is real commitment to improving the 'human' quality the likelihood of anything done in this area is extremely small. In many languages, including English, the word 'yes' does not necessarily mean commitment. It can mean 'I hear you,' 'I understand you,' 'I am saying yes, so that you leave me alone,' 'I say yes, not because I want to but because everyone else is saying yes.'

The ability to commit to something is a combination of both emotional and spiritual intelligence. It has an emotional component because it is not possible to commit to something unless there is a good feeling about the decision. It is spiritual because commitment is usually spurred by some deep-seated value or principle. It is also a state of being, in the sense that an individual or group makes a promise they say they will keep. There is also an assumption that with commitment goes integrity.

Commitment is not always easy, particularly in a diverse group which has conflicting interests. In groups where there is real commitment, there is often a shared vision, shared values and a common sense of purpose. There are also policies about social practices that every member believes in, including limits on what a board or executive committee will allow.

Commitment at this level takes time to establish. It also takes courage, respect and trust for people to share their deep beliefs and values with each other. One non-executive complained about going away for two days to discuss values. 'I thought it was a total waste of time. But now, I can tell you that it was one of the most rewarding things I have done in my life.'

The reason for this is that, in the right environment and with the right leader, it is an opportunity for individual board members to get in touch with what is really meaningful to them. If the human face of corporate governance is important, as I believe it to be, then the ability to know and speak one's mind is a starting point.

Knowing what is right can be a challenge. This is because we are often confronted with conflicting demands and paradoxes. One CEO asked me, 'How do I reconcile the need to support my family with my need to stand up to a tyrannical chairman and lose my job?'

Follow through to results

Once there is commitment to a decision, effective follow-up is essential.

The best results are achieved when those involved are clear about their practical commitments and timing. The better the support for implementing the decisions, the better the outcome. If a truly effective way of monitoring results is established and progress is thereby measured, there is a greater chance that successful results will be achieved. This is usually implemented thoroughly with regard to financial criteria, but seldom when it comes to the human side.

All this sounds self-evident. However, there are numerous occasions when good decisions fail and people suffer simply because of lack of follow-up. This not only applies to results that are profitable. It also applies to results that take account of the impact of decisions on human beings.

UNDERSTAND NEW CONTENT

Once a decision has been made and the results obtained, the group moves forward to the next challenge with its content and its process and the cycle begins again.

SUMMARY AND CONCLUSIONS

Corporate governance is mainly associated with the production of financial results, mainly in the interests of shareholders. As we go forward into the new millennium, I have noticed that more corporate governors are increasingly taking responsibility for the human side. This includes the quality of the working relationships between themselves and members of other boards.

Hopefully, whenever a crucial collective decision is made, it will be arrived at with the following questions prominently in mind:

- How can the decisions we make be of maximum benefit to the largest number of people?
- Are we spending enough time exploring ways to make a more positive contribution to the quality of human lives?
- How have the decisions we have made contributed to the well-being of people, and not only of the shareholders and ourselves?

Better decision-making skills lead to more effective results and not just only financial ones. Many investors, boards and executive committees are making a positive effort, yet much more could be achieved.

In this chapter a model for successful collective decision-making has been presented. The following chapter suggests practical guidelines for those who wish to improve their own contributions and the performance of the group.

Note
1. Hampel (1998) *Hampel Report on Corporate Governance*, Gee Publishing.

8 Guidelines for Interpersonal Governance

▶ ▶ ▶   EXECUTIVE SUMMARY   ◀ ◀ ◀

This chapter uses the framework outlined in the previous one to offer guidelines for effective interpersonal governance. In my experience, the majority of governance groups operate at well below their potential, and attention given to any of the areas discussed in this chapter could significantly improve performance. After outlining the criteria for improving performance, this chapter suggests ways to increase the understanding of the subject matter of the decision (content). This is followed by suggestions for planning the decision process, and then improving the process itself. Consideration is then given to the developmental needs of the governing group in the context of continuous change – offering guidelines on succession planning and selection and integration of new members.

I believe that most governing groups are only operating at not more than 50 to 60 per cent of their potential, not only in terms of the quality of decision-making, but also in terms of the speed and timeliness of decisions.

In the course of 20 years, and working with a large number of companies, I can count the number of outstanding groups I have found on the fingers of one hand. The others mostly function in an adequate way: after some discussion of the agenda, agreement is reached, a decision is made and the group moves to the next topic. Where there is a dominant leader, he or she usually makes the decision, and the group follows along. Sometimes the final decision is an excellent one, however it is reached.

But the point is that an adequate decision is no longer good enough today in view of the complexity of the challenges facing business leaders. It is increasingly important that all key players contribute, and integrate their contributions as much as possible. More effective deployment of resources, energy and talent also saves time – a commodity much needed.

In this chapter I focus mainly on interpersonal governance within the context of group meetings, although much of what I have to say applies equally to informal interactions between people, and relationships between groups of people. The sort of questions I am most often asked by chairmen and CEOs include the following:

- How can I improve the way I manage my people?
- How can I get the best out of them?
- Are we doing as well as we might?
- How do I deal with people who are not pulling their weight?
- What do I do about people who disrupt meetings?

Common questions from directors, either executive or non-executive, are as follows:

- What do I do with a leader who doesn't listen?
- How can I make decisions on late and inadequate information?
- How can I do something worthwhile with extremely limited time?
- Am I contributing as much as I could?

In answer to these questions, the suggestions in this chapter stem from many years of observing what works and what does not. Most of the advice is simply what good leaders and directors do naturally, often intuitively as 'common sense.' Other suggestions come from tools and techniques developed specifically to address these issues. Some of them depart from conventional practice, and require special training and different skill sets. Others amount simply to doing familiar things better.

The problem is that they demand an initial input of time and effort as well as unlearning outdated habits. Above all, they require commitment to develop, experience and experiment, and this may not be easy in view of the pressure to meet existing challenges and deadlines. It is always easier to leave things as they are, with the comfort of conforming to old routines and ways of working. But those who do take the time and energy to improve group performance will be rewarded, not only with excellent results, but they will also attract the very best people to their group.

CRITERIA FOR OUTSTANDING INTERPERSONAL GOVERNANCE

First, there is a checklist to help the reader estimate how well his or her group is performing. Rather than go through a long detailed list, the ten main qualities of outstanding groups are given, and I suggest you rank your group out of ten for each item. Then make notes about specific areas where you feel there is room for improvement.

Do not be disheartened if the descriptions seem to set impossibly high standards, as many of these criteria are already adequately met in a reasonably performing group. It is more important to focus on those areas that you feel need most attention. A chain is only as strong as its weakest link, and one small improvement in a vulnerable area can have a positive knock-on effect on other areas. For example, if there is better initial understanding of content, more time can be spent fruitfully discussing the issues.

Anyone can ask these questions for themselves, but for an objective review of his or her board or committee, it is more effective to ask for an external opinion. Most corporate governors lack the training and professionalism to do everything well. This is especially so when it comes to acquiring new skills, knowledge or expertise.

CRITERIA FOR OUTSTANDING PERSONAL GOVERNANCE

It might be useful to give each point a total score of ten and then rate how well you think your group operates.

1. All members of the group are of a high caliber, not only in terms of their competence to do the job, but also in terms of integrity and attitudes to people.
2. The best people are attracted because they are inspired by the values of the group.
3. There is a sense of enjoyment, camaraderie, expectation and readiness to tackle challenging tasks. The atmosphere is positive, and people are awake, alert and motivated to find the best solutions possible.
4. Levels of understanding of the topic are uniformly high. People have done their homework well in advance, and have a clear grasp of the key issues and what is expected of them.
5. Standards are high in all areas:
   Physical – financial competence, people are happy about their remuneration, share options and so on. The physical environment is pleasant and conducive to good work. Appropriate technology is available for providing information, and for exploring ideas logically and creatively.
   Mental – There are high standards of intellectual rigor, with stimulating and useful debate.
   Emotional – All members are emotionally mature. There is respect, trust, openness and honesty. Where appropriate, people are good at understanding others' feelings and expressing their own. When there are emotional difficulties, they are effectively dealt with – as far as possible at the time.

Spiritual – The team knows that there are shared and owned values and clear policies about their treatment of people. These include the way they treat each other, shareholders and other groups. They are concerned about the human implications of their decisions.

State of being – There is a powerful cohesive group that creates a positive image and impact. During meetings, there is constant awareness of individual and group states of being, for example mood and morale. If the energy or mood is low, it is improved as soon as possible.

6. All of these contribute to effective and appropriate ways of making decisions. The quality and timeliness of decisions are good. There is either authentic commitment or effective ways of recognizing and dealing with lack of commitment. There is a track record of exceptionally good results, both tangible and social. This is reflected in keeping shareholders consistently happy.

7. Members of the group know each other well enough to understand where they are coming from, their key contributions and pet hates so that they know how best to support each other. They also know who and when to approach if they need support from others.

8. There is willingness to understand and openness to learn and improve continuously – whether learning from each other, executives or outsiders. Personal and group development is part of the culture of the group. People take serious responsibility for their own and the group's development. This includes the wisdom to use consultants intelligently.

9. Everyone in the group feels that their time is well spent.

10. The structure and composition of the group reflects what is needed to fulfill present and future requirements. This means continuous upgrading of standards and people. The group has the right balance of people, both in terms of expertise needed and decision-making skills. There is good succession planning, selection procedures and integration of new members. Old members know when to go; resign gracefully and are optimistic about their next stage of life.

UNDERSTANDING CONTENT

The importance of understanding content has already been stressed. Whether it is because the content is too technical, the information is badly presented or communicated, or because there is simply not enough time to assimilate it, a board member who does not understand the issues is either not going to contribute usefully or else will hold up the meeting while details are explained.

The guidelines which follow should help to get the meeting off to a good start, with well-informed participants.

The content needs to include the human side

If the group is seriously committed to improving the human face they present, then any content, including financial reports, audits and the annual report, needs to address human issues. These must be given due importance rather than being tacked on as an embellishment because it 'does not look good' to leave it out. Annual reports need to have a dedicated section addressing the human implications of any future plans or decisions that have been made. Audit committees should always include this topic in their agenda. Indeed, many boards nowadays commission a social and ethical audit to be undertaken at the same time as the financial audit.

- Publicly announced visions, values and mission statements can be worthwhile if people genuinely believe in and live by them. It takes skill to create statements of shared value, and true ones provide a good foundation. Beware, however, of pseudo statements, or members of the group not living up to them. Failed promises produce lack of trust and diminished respect. People easily see through hypocrisy and deception. It is better not to have them at all.
- It is, however, important to have clearly understood and agreed policies and practices concerning the human side. These should be reflected in any document put before the group, or presented to other groups. These might include policies on the way people are treated, either individually or as part of a group; and on what sort of relationship the group wants with shareholders, major investors, the media, governments, pressure groups and others. There also needs to be consideration given to the implications that any decisions may have on the well-being of communities affected by them – not just environmental issues, but also the social implications for communities which might lose their livelihoods. Any group lacking clear views and policies on such issues cannot be seriously interested in the human aspects of governance. This includes investors.

Guidelines for improving understanding

- Before information is presented at a meeting, the presenter should be briefed as to what kind of information is expected, and the way it should be presented. Written guidelines may be helpful. People normally prefer a short summary, followed by a list of key issues and the way they are supposed to

deal with them. It helps those presenting material to have access to guidance
on the length, format and level of detail required.

- The chairman and non-executive directors need to feel they can obtain
  required information on demand. On the other hand, executives need to feel
  that this right does not permit non-executives to interfere in their
  management (unless there is a crisis or other legitimate reason).
- Some people can better assimilate information verbally than by reading it.
  Such people prefer to be briefed ahead of the meeting by someone
  knowledgeable about the subject.
- In some cases, the chairman and non-executive directors can be consulted
  before a presentation is finalized, and this will provide better value and
  additional information.
- People prefer to address actual and potential problems or drawbacks as well
  as being given the positive news. 'Gloss does not help us make decisions.'
- There are many different ways of presenting information. Endless
  PowerPoint, overheads or flip charts eventually become boring, and may not
  even be appropriate to the subject matter. Other forms of presentation may
  be more effective.
- There is no guarantee that everyone has the same understanding of content.
  If this is important, it would help to ascertain what each member understands
  to be the key issues of a paper or presentation. The earlier this is done, the
  less time wasted later.
- Too much or too little information can be a power ploy used by people
  wanting to rush through proposals without discussion. Remind them of the
  purpose of the group. If the ploy implies lack of respect for the quality of the
  directors, address the problem.

ENJOY THE PROCESS

Almost all the people I have worked with or talked to said that one of their
goals was to have fun at work. If someone is not enjoying the process of
governing most of the time, it is time to change the process, replace the people,
or resign.

DESIGNING THE PROCESS

Many people run, or take part in, meetings according to the same formula year
after year. Repetition and familiarity may enable things to run smoothly.

However, meetings and projects are often significantly improved if thought is given to tailoring the process to suit a specific meeting or series of meetings.

There are a number of points to be considered when designing the process. If they are not already in place they may take time to set up and use, but once that is done, they can significantly improve the quality of the work of the board or the executive committee. Three aspects are considered:

- Clarify the purpose of the meeting and match the process to it.
- Understand the group's potential for the task.
- Incorporate the above to design a time-efficient process.

Clarify the purpose of the meeting and match the process to it

Designing the process can only begin once there is a clear grasp of the purpose. The leader also needs to ensure that the right information or content is available well in time, and that it is sufficiently understood, as explained above.

Once this is done, it is possible to decide what is needed to produce the best outcome. This includes an understanding of the kinds and levels of intelligence required. It also helps to consider what stage a decision may be at, and what stage it needs to progress to. The Action Profiling Model (1) is one way to analyze the stages of decision-making. However, any good model will serve the same purpose. Given that this book is about the human face of corporate governance, it is suggested that the process leader is clear about his or her own values and standards of behavior. To help with this stage, the following questions have been formulated.

1. Are you clear about your core human values and what is acceptable and unacceptable behavior?
2. What is the ultimate objective of this decision?
3. What is the time-scale in which the decision needs to be made?
4. What is the nature of the decision to be made? What is needed for a really good decision?
5. What kind of thinking and intelligence is needed for the purpose to be fulfilled?

Physical
- Financial.
- In terms of material assets or physical activities.
- Physical presence and alertness of the group.
- Financial and material consequences for people and communities affected by the decision.

Mental
- What level of conceptual or intellectual ability is required for the decision?
- Does it require perspective of the whole? Or other kinds of perspective?
- Does it require creativity, analysis, or the ability to anticipate risks or gaps in thinking?

Emotional
- Does the group need to feel it has a shared sense of purpose?
- How united does the group feel?
- Have conflicts of interest been amicably resolved?

Spiritual
- Are the ethics, values and principles clear enough around the subject? Do they need to be defined?
- How much time do the moral issues need on the agenda?

Social
- What levels of cooperation are required between group members and other significant people or groups?
- What kind of communication will be necessary at the end of the meeting/project, how and to whom?

6. At the end of the meeting, how far through the decision process do you aim to be? Are you merely sharing information or options, arriving at a group decision, or deciding how to implement the decision in practice? In terms of Action Profiling, these stages are:
   - The Attention Stage – getting the right information or sharing the scope of possibilities.
   - The Intention Stage – arriving at a decision through building a firm case or clarifying issues, comparing options and prioritizing.
   - The Commitment Stage – committing to action. This involves deciding on the timing of an action, when to seize opportunities, speed up or slow down. It also includes anticipating consequences, seeing long-term goals or visions as well as planning stages of action and monitoring outcomes.

7. What kinds of thinking and exploring are needed? Logical, rational thinking and analysis? Creativity? Discussions around motivation and values? Anticipating consequences? Some combination of these? In view of this, what approaches, techniques or methods may need to be employed? How long will each of them need?

8. What types of expertise, knowledge or experience are required? Are they available in the group or do they need to be found?

Understand the group's potential for the task

The above questions are specifically related to the capacities required for the task. Although most chairmen/women and CEOs know their people well, they are often surprised at the additional potential that is available if they take the effort to find out. As most groups are operating well below their real potential, it is useful to find out what they are really capable of.

There are a number of ways of evaluating a group's true potential. Some of these you can do yourself, and others would be better done by a specialist consultant who is more objective. If you choose a professional consultant, be careful that they understand your business and are experienced and capable of dealing with directors without being prescriptive. It is also helpful if your values and theirs are compatible.

There are two levels at which you can consider potential. The first is that of individual contributions to the group. The second is to consider the combined potential of the group as a whole. This is harder, but more relevant to the development needs of the group. A starting point is to reflect consciously on current group dynamics, and the way this is likely to affect your design of the process.

How can each individual in the group give of their best?

The most effective ways of understanding individual potential are to ask and observe. In Chapter 5, Guidelines to Personal Governance, several ways in which an individual could gain greater self-knowledge were suggested. If the individuals have already gone down that route, the answers to the following questions may be of help. In any case, many chairmen and CEOs have found that asking the following questions is a useful guide.

QUESTIONNAIRE FOR GROUP MEMBERS

1. Briefly describe what you consider to be your true contribution.
   - In terms of your function, knowledge and expertise.
   - In terms of your contribution to the dynamics of the group.
   - In terms of what you contribute to decision-making.
2. What do you consider your area of outstanding ability? Think of three occasions when you really excelled. List the qualities demonstrated, and look for key qualities that were common to each occasion.
3. List what you consider to be your main strengths.
4. List what you consider to be your main weaknesses or areas of low motivation.

▶

5. Briefly describe the way you prefer to make decisions, for example, whether you need to explore ideas with people, or go away and think; whether you work intuitively, or need detailed information to analyze etc.

6. Briefly describe the ways in which you prefer to work; for example, do you like a lot of time, or do you work well under pressure?

7. Briefly describe how you prefer not to work.

8. How do you best like to receive information? In writing, verbally, a mixture of them, diagrams or other ways?
   How much information do you like to receive?
   How do you like it presented? For example, in the form of an executive summary, detailed reports etc.

9. What is your vision of the perfect board/committee meeting?

10. What would you like to contribute to the group that you are not already doing?

11. Briefly list your key values and principles related to corporate governance.

12. What contributions do you and could you make to improving the human side?

These questions will give some idea of whether there is potential waiting to be tapped, particularly if your directors are willing to be honest. It is, however, usually more effective to bring in an outside professional who not only goes through such questions privately with each member, but asks them additional questions in total confidence, such as:

- What do you think of the way you are given information, ideas and presentations?
- What are your current views of the performance of the board/committee?
- What does the group do well?
- What works?
- What does the group not do well?
- What does not work?
- How would you like to see it improved?
- What do you think about each board member?
- What for you are the key priorities of the business?
- What do you think about the group's principles and behavior as regards the human side? What 'face' do you as a group project to the outside world?
- How do these compare with your own values, beliefs and principles?
- How happy are you about the way in which you and other board members are treated?

Answers to questions like these, given to an independent outsider in confidence, often give a greater picture of the added value that directors can give. I have suggested above that one should not only ask questions, but also observe. My experience is that where people display the greatest passion or frustration is usually the area where the biggest potential contributions can be made. For example, one director was frustrated because his group never made decisions. 'It drove me mad.' The chairman ruefully admitted, 'We get carried away by the subject, and start opening up new areas to explore when the meeting should close.' So he asked the director to help bring meetings to a close, and remind the group that a decision was needed so it could move on.

These interviews can be written up in a personal report that each individual then endorses – with amendments as necessary – as an accurate description of their characteristics. The information can also be presented as a non-ascribed report that goes to the chairman or CEO. Either way, it can give them useful feedback about how the group could improve. With boards, such a report can form the basis of a 'Board Capital Report' that goes to investors. It enables them to understand the capacities of the board, and what is being done to deal with potential weaknesses.

A greater sense of what individuals can contribute provides a pool of untapped resources which can be made available in the right circumstances. A good profile provides a vital knowledge base, particularly if it is combined with other tests or personality descriptions. Then, every two years, it is useful to invite each person to review what he or she wrote, and ask whether they would like to update the information.

Once the leader has gained a clear idea of how each individual can add further value to their role, he or she is in a good position to plan how best to match these contributions to the decision-making needs.

Understanding group dynamics

It is good to have an idea of individuals' potential contributions, but if the board is not functioning well and working relationships are poor, much of this could be wasted.

Every leader who has worked with their group is likely to have some sense of understanding of how the group is operating, how much time is used productively and how much is wasted. Chapter 9 expands upon the themes of group disruption and wastage of time through dysfunctional behavior, but the following exercise could prove quite illuminating. If you feel you need more evidence, or want to check your perceptions, ask someone else to do the exercise and give you feedback.

It is simply a matter of precise observation: writing in columns with the headings Time, Behavior, Comment – as in the following example.

TIME	BEHAVIOR	COMMENT
9.15	Joe and Jim arrive 15 minutes late.	By the time everyone has said hello, 30 minutes wasted.
9.45	Begin the agenda. Check that everyone has read the minutes and got the material.	Twenty minutes spent going over the content before we could even begin to discuss it.
	Adam says he left his in Hong Kong. He clearly has not prepared for the meeting. Two others look as if they too are not prepared.	
10.05	Discussion begins.	Over an hour wasted.

LEARNING FROM THE ABOVE

1. Do not assume everyone has done their homework or understands the content. Get the CEO to check in advance and ask if anything needs clarification.
2. Remind people what was discussed last time, and what needs to be achieved in this meeting.
3. Let people know that if they do not arrive on time, the meeting will begin without them.

A number of suggestions and tools have been offered to gain more information about individual contributions and group dynamics, but there are many more techniques and methods available. Rather than get bogged down in technique, remember the main principle: whatever methods are chosen, they should take into account the following key questions:

- What does the group do well that serves the purpose of this decision?
- What does the group not do well that does not serve the purpose of this decision?
- How do the capabilities of the group match the requirements of the decision or task?

An example of the difference between a process that has design, and a process without the necessary design, is shown below.

EXAMPLE OF A CHAIRMAN PLANNING FOR A MAJOR ACQUISITION

CURRENT APPROACH

Purpose – To agree on whether the major acquisition of X should go through.

Process – To ask the CEO and finance director to do a presentation to the board.

Process – To ask for any views.

POTENTIAL IMPROVEMENT

Purpose – To agree on whether major acquisition of X should go through or whether there is a better alternative.

STAGE OF DECISION PROCESS

Attention – paying attention to the facts, and intention – arriving at a decision if possible.

KINDS OF INTELLIGENCE NEEDED

Perspective

Does it fit into our overall strategy?

Physical

What are the financial implications?

How feasible is this acquisition? How much time will it take to yield returns? (Remember the last expensive mistake.)

Mental

Ask John to evaluate how much a priority this acquisition is compared to others.

Ask if there are any other options.

Ask Hank to see if he can find any gaps in the presentation. Where are the risks?

Ask Tony about the cultural implications.

Emotional

The team needs to feel really enthusiastic about this one.

The conflict between Tony and the CEO needs to be resolved, or discussion will not be fruitful.

Spiritual

Are there any ethical implications?

If we said yes, how would the executives treat people from that company?

State of being

To move the board from being passive to being fully awake and involved.

METHODS

1. Ask the CEO to send a first draft (not detailed) to all board members a week ahead of time. Also, ask the CEO if there is any particular expertise or thinking he would like from the board.

2. Directors asked to submit questions they would like to put to me.
3. Key questions forwarded to CEO to be taken account of in his report.
4. At beginning of meeting, share intentions and the state of being desired.
5. Quickly check with people using a score out of ten of how enthusiastic they feel about the proposal.
6. Ask to lead a creative brainstorming session to explore alternatives. (Be aware that the board does not often ask 'what else') (15 minutes).
7. Have a general discussion (30 minutes).
8. Summarize discussion and check with board.
9. Check for levels of enthusiasm. Note concerns.
10. Next steps.

LEADING THE ACTUAL PROCESS

Every leader knows that a planned process may not always work in practice. People are unpredictable. It is a learning experience, and the art is to juggle what needs to be done with what actually happens. It is often a matter of needing to be flexible in order to give direction, but not so flexible that the direction gets lost and time is wasted. To run an effective process, it may be helpful to consider some principles and tips that others have found helpful.

The shape of the meeting or process

Goals

- The process produces maximum results for a minimum of effort and time.
- The meeting or project serves its purpose, and people go away happy about the way the process was run.

Principles

- The design of a meeting or project relates to what needs to be achieved.
- The most appropriate time-effective methods are found for different kinds of topics.
- Be prepared to change according to the capabilities of the group at the time.

Tips

- Expand your repertoire of ways of running meetings to cover a range of styles. For example, creative, rational, intuitive or ways of inspiring and motivating people.
- Explain to the group your intentions for the meeting. If you feel you have to change from the original format, explain this as you do so.

- Allow more than enough time to complete the project or meeting successfully.
- Allow time at the end for a review of the process.

Physical

Goals

- To ensure that people are awake and physically energized.
- Also, that they are happy with the physical environment they are in.
- To provide the technology necessary for sharing information effectively.

Principles

- Make sure that the physical environment and energy level encourages good work.

Tips

- Check whether the physical environment is comfortable and pleasant and that you can see everyone at the same time. Think about going to different venues for different kinds of thinking.
- Make sure in advance that information and presentations are up to standard.
- If people are likely to be suffering from jet lag, provide quiet places for them to doze for a few minutes if they find their attention is lapsing.
- If you are adventurous, try music and creative techniques to liven the place up, and to enable people to shift to other levels of awareness.
- Consider using physical objects such as maps, models, products, figures and so on to gain perspective and generate ideas.
- Make sure that meals during meetings are light and not sleep-inducing.
- Take breaks between items and when energy is low.

Mental

Goals

- Intellectual rigor.
- High-level debate.

Principles

- Keep thinking at the right levels for the task.
- Maintain the right level between perspective and detail, and between conceptual and practical.
- Ensure that appropriate modes of thinking are used. See Figure 5.2.

Tips

- Ensure sufficient people of a high intellectual caliber understand the issues, get to the core and think logically and rationally.
- Check that people understand the purpose and content before a debate.
- It helps to remind people what happened at the last meeting.
- It can save time to provide a short summary of content, and a list of key issues that need to be addressed.
- Remember that questions often provoke better responses than statements.
- Focus on gaining a perspective of the whole.
- Keep away from too much detail unless it is absolutely necessary.
- Encourage people to listen first and gain understanding before responding.
- Check that there is shared understanding before moving on.

Emotional

Goals

- To establish and maintain high motivation and morale, including trust, honesty and openness.
- To deal with emotional issues in real time, or as soon as possible.
- To develop high levels of emotional sensitivity, competence and maturity.
- This includes the ability to give and receive love and compassionate detachment.

Principles

- Encourage an atmosphere of trust and respect; acknowledge openness and honesty, including emotional honesty.
- Emotional safety enables emotional problems to be resolved and not cause the group to dysfunction.
- Blaming seldom achieves the desired results.

Tips

- Assume good intention.
- Do not take responsibility for other people's emotions.
- If someone looks distressed or unhappy, check it out with them. Your original interpretation may not be correct.
- Learn to understand and express your own feelings when appropriate. Encourage others to do the same constructively, and in a safe environment.
- If you unintentionally threaten, criticize or undermine people's confidence, a timely apology can save a great deal of pain, distrust and antagonism.
- Be aware that intellectual explanations, solutions and arguments are not

appropriate for emotional problems. Respect the way people feel. Ask what they need in order for their emotional needs to be met.

- When dealing with emotions, use emotional language such as 'happy, contented, satisfied, fear, anxiety, concern, sad, angry' and so on.
- If people have difficulties with others, suggest that they share their problems directly with the other person, possibly with a facilitator. Talking behind backs causes more difficulties than taking the courage to confront.
- Do not tolerate emotionally disruptive behavior. If someone is causing distress in the group, take them aside privately and sort it out. If behavior indicates that there are emotional issues, stop the process, get the group back to a good state of being and continue. If there is a negatively, emotionally charged atmosphere, productivity will be low.

Intuition

Goals

- To use intuition either to initiate ideas or perceptions or to check out whether things are going well or wrong.

Principles

- Always respect your own and other people's intuition.

Tips

- Always check intuition against facts.

Spiritual

Goals

- To have clear values, principles and policies that are lived in practice by all board or committee members.
- To establish and maintain a reputation for integrity.
- For the board to commit to improving the human as well as the financial side.
- To have a consistently good state of being.

Principles

- A positive state of being has a positive impact.
- Clear policies and principles add value to the reputation of the group.
- It is not necessary to be perfect to be valued as a human being.

- Least damage is done when people understand and support each other to limit weaknesses.
- Integrity is the cornerstone of all behavior.

Tips

- Check that you are clear and happy about your own values, beliefs and principles and what is and is not acceptable behavior.
- Make time and use appropriate methods to enable the group as a whole to do the same.
- Ensure that once agreement has been reached concerning shared values, there is consistency of behavior. Remind people when and if they slip. Ask them to remind you.
- As a leader, ensure that your state of being is calm and positive and that you feel powerful in a positive way.
- As soon as you feel the state of being is not good, stop the process, take a break or do whatever is necessary to bring it back to what is needed to fulfil the purpose.
- If you feel anxious or insecure, try to get back to a positive state of being. Remember to breathe properly and to relax.
- If there are members who are diametrically opposed to the values and culture of the group, they should be asked to leave. This is not the same as differences of opinion, which should be encouraged.
- Find a mentor or mentors to help support you keep to your own spiritual path and to deal with paradoxes.

Conflicts of interest

Goals

- For conflicts of interest to be satisfactorily resolved, both for the good of the group and for the individuals concerned.
- If this is not possible, to limit any potential problems.

Principles

- Conflicts of interest should be recognized, discussed and consciously resolved.
- No individual should allow their personal interests to undermine the group, the shareholders or the corporations that they are governing. If this cannot be resolved, it would be better for them to leave.
- If people are stuck, they should employ an objective external arbitrator or moderator.

Tips

- If you personally have a conflict of interest, write down the pros and cons of each interest. List the implications of each route. Then ascertain what your real truth is, and whether you are acting out of integrity. If you are, then go for it.
- Regularly check members to see if they have conflicts of interest. If relevant, help them resolve them.
- If you see a board member is looking uncomfortable, it may be because they have a conflict of interest. Check it out as soon as possible.

Communication

Goals

- High quality information is communicated on time and understood.
- In the group, there is mutual understanding of, and building upon, different points of view.
- People know the appropriate ways and times to communicate, and do this well.

Principles

- Attention should be paid to the right amount of information for the purpose.
- Communication skills are essential for good corporate governance at all levels – with the board, the executive, the shareholders, investors and other stakeholders.
- Bad news and potential difficulties need to be communicated earlier rather than later.

Tips

- At the beginning of each meeting, communicate clearly the purpose and desired outcome.
- Once you have set up expectations, do not change direction unless absolutely necessary.
- Silence and time to reflect can be as valuable as talking.
- There needs to be freedom for people to learn and communicate in ways that work for them.
- Encourage listening for understanding before discussion.
- It helps for people to know what kind of communication is required of them.

Periodic reviews

Goals

- An effective board or committee that is capable of, and committed to, continuous improvement.

Principles

- The performance of the board or committee needs regular reviews to establish how effective they have been, and how effective they need to be in the future.

Tips

- Build into the corporate timetable enough time to review the performance of the group at least twice a year.
  1. What has worked?
  2. What has not worked?
  3. What needs to be improved and how?
  4. What have we achieved in terms of the human side?
  5. What have we achieved in terms of our own working relationships?
  6. What have we achieved in terms of relationships with others?
  7. What have we each personally achieved?
  8. What have we achieved in terms of the corporations being governed?
  9. What have we achieved in terms of the communities impacted by our decisions?
  10. What else have we achieved?
  11. Have we reached our goals?
  12. What are our goals for the future?
- Regularly bring in professional observers to give objective feedback and recommendations.
- When a new member comes onto the board or the committee, ask for a written report on their first impressions and what they think are the board's or committee's main strengths and needs for improvement. Catch them before they have become integrated into the group.

GROUP DEVELOPMENT

Every leader knows that managing the process of leading a group of people in corporate governance is a learning experience. If it stagnates, the group will be in danger, because a board or committee is only as effective as its capability to meet current and future challenges. Continuing to work in the same old ways will result in being left behind. Any board or committee should, therefore, always be in transition. Figure 8.1 suggests a framework for considering what development the board or committee may need.

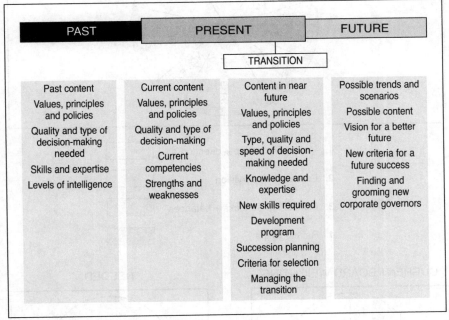

Fig. 8.1   A framework for progress

The type of leadership the board or committee may need in future is also important and may be different from the present.

The circle in Figure 8.2 shows the change in focus that one group thought it needed for the next stage of decision-making around the future of the corporation.

Tips

- List both current and future standards required for the board or the executive to do its job well.

- List ways in which you and the board or committee need to improve the dynamics of the group as well as the quality and speed of decision-making.

- Choose the key areas needed for development. What knowledge, training, or new skills are required? Set success criteria to measure progress.

INDIVIDUAL WEIGHTING

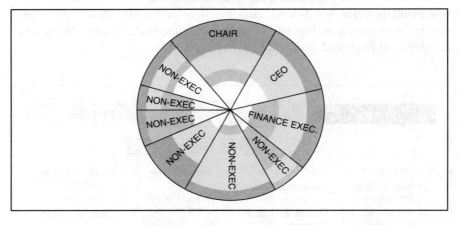

1. Vision, leadership + direction

2. Executive supervision

3. Monitoring – checks + balances

4. Relationships

CURRENT BOARD WEIGHTING   NEEDED

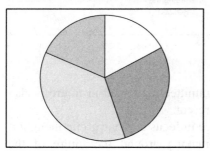 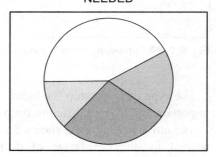

Fig. 8.2   Leadership/exec balance at board level (depends on present and future needs of the business)

- Ensure that there is commitment to develop, and to devote the necessary time to make improvements.
- Find a coach/mentor to ensure that learning occurs.
- Design a program over time in order to achieve these developments. (Most groups take between one and two years to realize their full potential,

depending on levels of commitment and time and the capacity of group members to learn.)

- Consciously set aside time, and regard this time as sacred rather than expendable.
- Build development into normal meeting times as a part of the agenda.
- Also, encourage the self-development of members and budget for it. Have external mentors or coaches available if members wish to pursue self-development, talk anything over or learn new skills.
- Regularly monitor progress.

SUCCESSION PLANNING, SELECTION AND INTEGRATION

The standards needed for the next stage of a board or committee will to some extent determine the structure and composition of the board or committee as it evolves (see Chapter 10). This will also be affected by current strategy, especially if there are mergers or acquisitions that involve new board members. Another consideration is when, and how many, board or committee members are likely to be retiring.

Succession, like remuneration, is often a difficult thing for boards to address. There are too many emotional ties and loyalties at stake.

Tips

- Have an annual review of what skills, levels of thinking, experience and expertise are currently required and may be needed in the future. List the criteria for evaluating future success.
- Check that the board has the capabilities to meet these needs. If not, consider whether the composition of the board could be enhanced to reflect what is needed.
- If any board members are resigning, use the above criteria as a basis for considering the expertise and qualities needed to replace them.
- Always have people ready to replace key people, either internal or external. Life is unpredictable, and if someone leaves and there is no good back-up, the group and the corporation could be left with a gap that could put the shareholders' money at risk.
- Check that there is adequate balance in the following areas:

    Legal and fiduciary.

    Remuneration.

Leadership and direction.
- Vision or business propositions.
- Strategy. Negotiations.
- Internal culture, values and principles.

Monitoring. Auditing.

Relationships, internal and external.
- Shareholders.
- Major investors.
- Regulatory or global bodies.
- The corporation.
- Stakeholders.
- Communities affected by decisions.

Knowledge and expertise in specific areas relevant to the corporation and requirements of shareholders.

Knowledge of the implications of information technology and other technical and scientific trends on the business and the industry.

- Types of intelligence and perspective needed:

Conceptual vs. practical.

Creative vs. rational.

Multi-dimensional perspectives of the whole.

Knowledge of how the parts need to fit together in practical terms to achieve results.

Physical
- Financial.
- Knowledge of how things get done.
- Physical environment.
- Understanding of the physical consequences of decisions on people.

Mental
- Intellectual rigor.
- Ability to analyze, explain and argue rationally and logically.
- Modes of thinking.
     Attending to information and options.
     Preference to share or go private.
     Committing to action.

Emotional
- Emotional maturity.
- Ability to be effective in sharing and dealing with emotions.
- Sensitivity to other's feelings.

Understanding of the emotional implications for people of decisions taken.
- Sense of humor. Able to diffuse tensions and conflicts.

Spiritual
- Integrity, compassionate, able to give and receive genuine love.
- Clear values and principles. Places value on the human side.
- Able to reflect on the deeper meaning of things.
- Positive state of being.

Communication/social
- Good at communicating in writing and verbally.
- Networker. Good contacts.
- Awareness of the need to communicate well with various parties.
- Good with the media.
- Good at presentations and addressing groups, both large and small.

LIFE CYCLE ISSUES

The composition of a group will reflect the life stages of the organization such as birth, infancy, adolescence, adulthood, maturity, old age, death, rebirth and regeneration. Different kinds of people understand and are better suited to govern a corporation at various stages. For example, someone who is good at turning around companies and who enjoys the creative challenge of beginning again in difficult circumstances is not usually as good at maintaining a well-established empire.

The life stage of the group itself may also need consideration. If a group has been going for a number of years with the same people, new blood may be needed. On the other hand, if the CEO and other board members are young, maturity, wisdom and experience are needed as a balance.

ACQUISITIONS AND MERGERS

Any major merger or acquisition is likely to change the life stage and nature of the corporation, and therefore the leadership composition. This is seldom sufficiently thought about in advance. Directors may join a board or executive board/committee purely as a consequence of a merger or an acquisition. Unless this is well managed, it can cause enormous difficulties and disruption, particularly if there are power games or culture clashes.

Tips

- Check that culture, values and beliefs are compatible with the existing board or committee. If they are significantly different, consider the implications well in advance. If necessary, reconsider the deal.
- Do a lot of work before directors come onto the board to explain the culture, and what is expected of directors, and co-create the best ways in which integration of the companies can take place.
- Remember that it is natural for someone from another company to want to protect the interests of that company rather than the one they are joining. This needs to be discussed in advance in an honest and constructive way so that the needs of both can be met.
- A well-thought-out induction program is always useful.
- Existing members need to consciously consider the best ways of welcoming newcomers, and integrating them into the group. If this is done in a caring way, it makes a great deal of difference.

Integration of new members

Any changes on the board – whether from mergers and acquisitions, additions or replacements – will involve adjustments on the part of the group. If the group as a whole participates in deciding the best ways of integrating new members, the change tends to be made more quickly and more smoothly.

Different people take different lengths of time to fully understand the complexities of the job and of the corporations. In the case of executive directors, it should take not more than a year, though many are expected to achieve results in a far shorter time. In the case of non-executive directors, it can take up to two years to go through the learning curve.

Tips

- Find out from the newcomer what they need in order to reach an understanding of the organization and key governance issues as soon as possible. Each person is likely to have a different approach and to learn in different ways.
- Regularly update your induction program.
- Discuss with the existing directors how they would best like to welcome and integrate new members.
- Regularly check that a person is progressing as well as they should. Find out what additional information and support they may need.

RETIRING WITH GRACE AND DIGNITY

Finally, but not least, is the smooth transition from the old to the new. For the people concerned this can be either an extremely painful or a rewarding experience.

For investors, one of their main concerns is whether an existing chairman or CEO is going to resign when they should. If they leave without a successor being chosen in a reasonable time, a gap can be left which leaves the board and the corporation leaderless and possibly at risk. Share values go down and shareholders are not happy.

Many hold on to the bitter – and it often is bitter – end. Some of them are ruthlessly deposed. It is a terrible pity to end a spectacular career in an unhappy way, even if it is understandable. Few people want to let go of things when they are successful, have a familiar group of people to relate to, and which is a large part of their lives.

If the human side is important, then the way in which a person values himself or herself at the time of retirement, and the way in which he/she is treated, is extremely important. Succession is an excellent indication of how well this is being handled.

Tips

- If you are a leader, set criteria for yourself as to when you feel it is right for you to go. Discuss this with the board, and arrive at a mutually agreed decision about criteria and timing.
- It is your responsibility to mentor and train a successor.
- Remember that a reputation is built up over a lifetime. It can be lost and forgotten if parting becomes acrimonious.
- An outside counselor/mentor is especially useful at this stage. You can test ideas, talk through what you really want and explore alternatives about what to do next.
- Leaving a much-loved company and familiar faces is as much a loss as the loss of a family member or friend. Some people go through a bereavement process although they may not be aware that this is what is happening. If this can be talked through and understood, it can help a great deal. It is also important to find ways of establishing connections with like-minded people in some form or another after retirement.
- The nominations committee should think about back-ups and succession at least three years ahead of retirement, and insist that appropriate measures be taken.
- The person in charge of succession and development should always have a list of potential replacements, or new board members, as well as connections with good headhunters.

SUMMARY AND CONCLUSIONS

There are so many ways in which interpersonal governance can be improved that it has been difficult to know what to include and what to leave out. However, the main point of this chapter has been to indicate that there is more to interpersonal governance than running projects and meetings in the same old ways, year after year.

Much of what has been written here is already used instinctively by the best leaders. It does not, however, do any harm to check the extent to which one is conscious of, and fully understands and masters, the decision-making process and the quality of the dynamics of the group.

It is also important to use the services of an outsider from time to time to review the performance of the board or committee itself, and assess what is needed for improvement.

Note
1. Ramsden, P. and Zacharias, J. (1993) *Action Profiling*, Gower.

9 The Shadow Side of Corporate Governance

▶ ▶ ▶	EXECUTIVE SUMMARY	◀ ◀ ◀

This chapter deals with those aspects of human nature that cause unnecessary suffering to others. It is based on the assumption that much harm is done unconsciously. It is also based on the assumption that everyone has feet of clay, and has the potential for creating harm to others. However, if there were greater understanding, some people might be able to avoid certain mistakes, or at least, limit any damage.

Three areas are examined. The first is normal weaknesses and blind spots that a person or a group of people might have in terms of the roles they play. These are described under the heading 'Common Examples of Abuse of Power.' They include elements of a person's core nature, and aspects of intelligence that may not be fully developed. The second is an understanding of the damage caused by people trying to satisfy basic needs in unhealthy ways. Suggestions about how to deal with these are made. The third is an understanding of the damage that inappropriate psychological defenses can play in human interaction and thereby affect the quality of decision-making. Finally, a short set of guidelines is offered for those who want to avoid or mitigate the shadow side, if possible.

Scandals of one form or another are reported in the world's media every day. One organization violates regulations and causes severe pollution and environmental damage in order to cut costs. A chairman resigns because of insider dealing on his board. A group of directors commits fraud, or important information is withheld. Others secretly corrupt stakeholders and cream off funds for their own use until a company goes bankrupt. Senior figures are caught having sexual affairs with members of their staff. The lawyers have a wonderful time with litigation caused by revenge games.

Many other things go on that are less newsworthy but potentially just as harmful. For example, tyrannical chairmen and CEOs systematically break the moral self-respect and confidence of their staff. Directors manipulate others to sell their shares so that they can make a killing. The wrong decisions are made

147

because weak non-executive directors cannot stand up to a dominant chairman or CEO. Cartels hatch plots that stay just on the right side of the law.

There is a well-known saying that power corrupts and absolute power corrupts absolutely. In one sense that may be so but, in my experience, more people cause harm unintentionally. However, most powerful people do find constructive ways of channeling their energy without causing long-term damage.

Nonetheless, the list of abuses of power is endless and all too familiar. Even excellent board members with a reputation for integrity have been caught unawares and human nature, in its infinite complexity, is unpredictable. So this chapter looks at some of the possible causes in the hope that casting a light into these shadow areas may increase awareness and enable people to better manage their own and other people's weaknesses.

'GERMAN LEESON' COSTS FIRM £18M

Yesterday saw frenzied attempts by the German media to identify the person who squandered nearly £18m through rogue trading.

Like Leeson, the German employee of Electrolux gambled on futures and lost. As the situation deteriorated, more was gambled and lost spectacularly on the Frankfurt stock exchange. The person responsible was employed at Electrolux's own Frankfurt bank to manage liquidity, short-term borrowing and foreign exchange actions.

A company spokesman said, 'We can confirm that an employee of Electrolux's internal bank has engaged in a number of unauthorized and irregular trading contracts, resulting in losses of approximately £17m to £18m.'

Like Leeson, the perpetrator could face up to seven years in jail with the introduction of stiff new penalties for insider trading in Germany.

Source: *Yahoo!* Headline, 4 January 2000

US WARNS MCDONNELL DOUGLAS

The US Department of Justice is planning to indict McDonnell Douglas Corp. together with a Chinese aerospace company on charges of violating export control laws, unless the companies concerned agree to plead guilty to lesser charges in the coming weeks.

Source: *Wall Street Journal Europe,* 11 October 1999

THE ILL EFFECTS OF CHEMICAL DEPENDENCE

A major global petrochemical company's attempts to silence environmental critics have got it into deep water. The company has been operating in South America since 1960 and the government there has always allowed it considerable freedom in its operations because the economy is so dependent upon its substantial revenue. However, in recent years a local Indian tribe has joined hands with global environmental groups to protest at the alleged destruction of the fragile ecosystem which provides their livelihood. The company claims to have had no involvement in or knowledge of this year's brutal attacks upon the local community, insisting that it was a 'deep-seated inter-tribal problem.' However, a recent report by Greenpeace casts a very different light on the issue.

Story fabricated from composite sources, February 2000

Ever since I was a child in South Africa, I was brought up to honor human dignity by my parents and people such as the Nobel Prize winners Chief Luthuli and Nelson Mandela. I was also exposed to the effects of tyranny when my parents and friends were banned, tortured and imprisoned. Some died. What I learned is that it is not difficult for a group to display psychotic traits when abusive behavior is accepted or taken for granted. Not only that, but given the wrong conditions, almost everyone has it in them to be destructive. Indeed, as Oscar Wilde well knew, 'Everyone destroys the thing that most he loves.' (1) Perhaps because we are ruled by the cycle of life, birth, death, and rebirth – Eros and Thanatos – our capacity to be destructive is as natural as our capacity to be creative.

In Chapter 3, 'A Model for Self Governance,' it was suggested that a person's response to life is influenced by the extent to which their basic needs have been met. Two main themes are explored in this chapter. The first is what happens when basic human needs are not met. Here, I have been greatly influenced by the works of Manfred Max-Neef who suggests that certain ways in which people try to satisfy their needs are counterproductive. (2) The second theme is the way in which every individual in a group copes with two sometimes irreconcilable demands: the compulsion to satisfy individual desires, and the need to conform to the requirements of the group. Here, I have been inspired by the work of Guy Swanson who suggests that people build defense mechanisms that can be harmful both to themselves and the group. (3)

THE SEARCH FOR FULFILLMENT

Manfred Max-Neef suggested that there are certain basic needs that human beings try to satisfy. These include subsistence, protection, affection, understanding, participation, idleness, creation, identity and freedom. I have taken these into account but have incorporated them into a framework more relevant to corporate governance.

In Chapter 3 the concept of intelligence was widened to include the physical, mental, emotional and spiritual aspects. More advanced forms of intelligence, such as the capacity to give and receive love, wisdom, knowledge and creativity, integrate all four. The theory is that if an individual experienced severe deprivation in the past, whether it was real or simply imagined deprivation, similar experiences tend to trigger a neediness that they try to satisfy in oblique ways, some of which can be destructive. Table 9.1 lists needs, first under the four basic headings – physical, mental, emotional, and spiritual – and then the needs stemming from being part of a social group, the need to give and receive love, to be creative and the need to accept and be oneself.

If any of these basic needs is denied or repressed at crucial times of a person's life, it is highly probable that they will find vicarious and unhealthy ways of satisfying their needs. This often leads to destructive and dysfunctional behavior. In corporate governance, the most common forms of negative behavior come under three main headings: proactively aggressive or destructive; passive violence; and neurosis or psychopathological tendencies.

Before describing some of these, it is important to remember that business leaders possess a complex mixture of characteristics and that everyone has their strengths and values. For example, one bully that I knew had brilliant commercial ideas, which he followed through and made happen. He really cared for the company, and people both loved and hated him. They deeply respected his intuition and his creativity, although they hated his abusive behavior, because

---

UN CALL FOR POLICY CHANGES AT TRADE TALKS

The United Nations Secretary General, Koffi Annan, agreed that people should indeed be concerned about the environment, child labor, human rights, poverty and the commercial use of scientific and medical research. He expressed his belief that the real solutions lay in improved policies at national and international level, rather than trade restrictions that could aggravate poverty and obstruct development.

Source: *Financial Times*, 1 December 1999

his judgment and criticisms were often fundamentally right. It is also useful to point out that although some traits are exaggerated in certain people, most of us can slip into these modes of behavior in times of weakness or severe stress.

FRAMEWORKS FOR UNDERSTANDING THE SHADOW SIDE

Please note that the following frameworks and descriptions are an exploratory tool. The value lies in testing these ideas – I am not presenting an academic study of reality. This is better covered by psychiatrists.

How to use the frameworks

The intention is to approach an understanding of the shadow side in different ways. Because the shadow is by nature largely unconscious, it is not easy for anyone to become aware of his or her own shadow. In fact, most of us deny those aspects that we most fear in ourselves. The first step is to accept with compassion that every human being, including oneself, has a shadow and to let go any expectation that perfection is attainable. The second is that understanding the shadow is a continuous journey of exploration. Changing or mitigating negative or destructive behavior is often a lifetime struggle. Sometimes the skill is in damage limitation.

In corporate governance, the shadow side covers all four levels. At the systemic level, aspects of a whole system can have a negative impact on people. At the intergroup level, groups can sustain and condone corrupt practices. At the interpersonal level, the dynamics between people can be destructive as well as counter-productive. At the personal level, an individual is both affected by and influences their shadow side. Hopefully, the reader will use these tools to think about the implications at all levels.

At the end of the chapter some suggestions will be made for dealing with the shadow side or limiting its negative effects. These suggestions will be fairly basic, as there is more to be said about this topic than a book of this size can handle. Effective handling of the most difficult aspects requires painstaking skills training as well as specialist input. However, there are some guidelines that the reader can safely follow which can have beneficial effects.

Meanwhile, it is suggested that the reader studies the examples in this chapter and ascertains whether they recognize aspects of themselves in some of them. If the reader feels that he or she is perfect, I suggest they look carefully at delusions of grandeur! However, it is important to point out that this shadow side is the area of self-awareness most difficult for any individual to access on their own, and that it is often better done through feedback from other people. The reader might find it easier to recognize such behavior in others, or groups that they know. Finally, the examples are only indications, and it is better if the reader draws on their own experience.

Three areas are examined. They are as follows:

- Examples of abuse of power.
- Difficulties arising from deprivation of basic needs.
- Ineffective use of destructive defenses.

In each area, some specific recommendations are given. More general guidelines are given at the end of the chapter.

EXAMPLES OF ABUSE OF POWER

These are patterns of behavior that I have found most prevalent in working with chairmen, CEOs, non-executive directors and executive directors. They are also those most frequently mentioned by board members. I have placed them under three headings: overt, proactive abuse of power that is obvious; covert, devious abuse of power; and misuse of power through psychopathological tendencies.

Overt abuse of power

The bully

The most common is the bully. The bully dominates people, threatens them, emotionally undermines them and takes away their self-respect by constantly criticizing them. They apparently take delight in destroying weak people. They are usually unfamiliar with the problems of others and are insensitive to their feelings. The downside of bullies is that they breed a culture of fear, defensiveness and secrecy. Because the bully is dominant, he or she is usually surrounded by weak people, while the best often leave. They therefore make mistakes without being challenged. Worst of all, there will not be a suitable successor to take over. Bullies often suffer from severe feelings of inadequacy, insecurity and a need to control.

Delusions of grandeur

The second is a belief in one's own invincibility and superior judgment. As one chairman says, 'you encourage people to tell you what you want to hear.

Everybody likes to be told they are doing a great job.' This type of behavior drives people into isolation: again, they can make mistakes without being challenged. People who believe that they are invincible also suffer from complacency, and waste the expertise and knowledge around them. Sometimes the delusion is founded on genuine early success, but times or circumstances have changed.

Obsessive drive

These are people who are achievement-driven. They have to make more and more money, get to the top of the status tree and stay king for as long as

---

THE MD WHO WAS ALWAYS RIGHT

The managing director of a small computer hardware company had lost two successive marketing directors and was already at loggerheads with the latest incumbent about the company's shrinking media presence. The director wanted a major advertising promotion, but the MD dismissed that as a waste of money because the company had always got excellent coverage via cost-effective PR programs.

On the face of it the MD was simply a bully, always interfering with the marketing director and overruling his proposals – but this was not really his nature. He founded the company as a breakaway from a large corporation that made a whole department redundant. With strong socialist principles, he had made personal sacrifices to create the company and provide secure employment and a good working environment for his staff. The company flourished, partly because of his natural marketing and PR skills. And that was the problem; he had had such success running the marketing that he could not let go when the company grew and someone else was recruited to run the marketing.

Journalists told his marketing director that they had given the company a lot of free editorial coverage when it was a gadfly startup, but now the company was a niche market leader they would like to see a bit of advertising spend to 'oil the wheels.' But the MD would not allow it. 'No advertising – don't let them bully you,' he insisted.

It was only when the key industry journal sabotaged the company's biggest product launch by placing competitors' full-page ads facing the launch article that the MD faced the fact that times had changed. He admitted that perhaps he no longer knew all there was to know about marketing, and was at last able to let go and allow the director some freedom to run his own department.

possible. They are workaholics, obsessed by passions to the exclusion of everything else. This can undermine their judgment and relationships. Sometimes the obsession is utterly inappropriate to an actual situation – what Max-Neef calls a pseudo-satisfier, an attempt to satisfy a basic need such as love, affection or acknowledgment. The greater the battle, the less the satisfaction.

Overt internal competition

Another negative form of behavior is when people who are supposed to be co-operating in a group are so busy competing and trying to destroy each other that they forget to keep their eye on the ball. Meetings become battlegrounds where the most aggressive can score points. More energy is directed to internal competition than to competing in the marketplace.

Covert abuse of power

This is possibly the most difficult aspect to deal with because it lies hidden until it is often too late. Things are done through secrecy and deceit, often with everyone concerned seeming to be utterly honest and charming. Shakespeare knew all about this and excellent examples are found in his characters of Richard the Third and Iago in *Othello*.

Deceit

This is when someone deliberately misleads an individual or a group by obscuring facts, dubious intentions, or unethical, even criminal behavior. People might cover up for each other. They might pretend to be friendly when they are plotting to get rid of someone behind his or her back. When deceit is taking place, many people have an intuitive hunch that something is amiss, but often dismiss it for lack of evidence.

Secrecy

Withholding information or producing unintelligible information, any non-executive director will tell you, is a well-known form of indirect power play. Inadequate information makes it difficult to know whether the right decisions are being made. One director said that 'I feel as if I am expected to do my job with blindfolds on.'

Manipulation

This is where board members and executives are manipulated through flattery or vague threats or promises to do what a particular person or group wants. People who manipulate others tend to undermine people's ability to share their own views in a normal way. Manipulators also often enjoy intrigues and 'politics.'

'Politics'

This is the activity that goes on behind the scenes when ambitious people scheme ways of setting the stage for their own success, usually at the expense of others. They are likely to promote themselves with influential people, and at the same time denigrate their colleagues. People who play 'politics' can also be economical with the truth.

Psychopathological tendencies

Many of the most talented leaders and, indeed, artists display some form of psychological illness or neurosis. For example, people with manic tendencies can achieve outstanding success because they think they can do anything. On the other hand, many people suffer from temporary depression at some stage during their working career.

The problem with people suffering these illnesses is that they may have a distorted view of reality, which is difficult to detect and can be dangerous. For example, one chairman was paranoid, and almost convinced me that his board was plotting against him when it was not.

The most dangerous psychopathological tendency is socio-psychosis where a person has no moral or ethical sense and therefore believes that any form of behavior, even criminal behavior, is valid. They have no conscience and may be difficult to detect because they seem quite normal otherwise. Indeed, it can lead to remarkable success against competitors by removing constraints under which they operate.

EXAMPLE OF A SOCIOPATHIC DIRECTOR

In one company, a director who had previously sold his company to the group came on the group board. He won the votes of two other directors by doing them favors. He started creaming money from his old company. Other board members sensed that something was going wrong. By the time the difficulties emerged, the director had blamed the CEO and set it up so that he and the non-executive directors who were not involved in the scam were voted off the board. He took over as CEO. Within two years the company was bankrupt. However, he and his colleagues had stacked up considerable sums in their offshore bank accounts and got away with it scot-free.

DIFFICULTIES ARISING FROM DEPRIVATION OF BASIC NEEDS – ALL LEVELS, SYSTEMIC, INTERGROUP, INTERPERSONAL AND PERSONAL

In the course of my work, I discovered very early on that behind the problems I was helping to address there were often prior issues that had first to be resolved. People were often struggling with painful emotions that they did not know how to handle. I also discovered that if it was possible to recognize these fundamental needs, then the problem could be much more easily resolved. This seems to apply across the board, be it an individual, a small group or larger community.

In the mid-nineties I had the privilege of listening to Manfred Max-Neef talk about his work. I was excited by his descriptions of different kinds of satisfiers, and the fact that some things people try to satisfy basic needs do not succeed. In spite of the fact that his work was mainly with poor communities in Latin America, it became clear that many of the needs he described were basic to everyone, including those involved with corporate governance.

Some of the needs he mentioned have been incorporated into the model that follows. I have also related certain needs in terms of whether they are physical, mental, emotional or spiritual.

Table 9.1 includes lists of needs, the potential effects of deprivation (real or perceived) and possible outcomes if needs are not met. It is important to remember that perceived deprivation can have as powerful an effect as real deprivation – a rich child has the potential to feel as deprived as a poor child. An experience can trigger off a painful memory and a similar response even though it may happen years later. When we try to satisfy needs in ineffective ways, the less we achieve, the harder we try – but in the same old ineffective ways. Table 9.2 outlines difficulties arising from ineffective efforts to satisfy perceived or real needs.

GUIDELINES FOR DETERMINING NEEDS

Table 9.1 indicates some basic human needs and the potential effects of deprivation. They apply at all levels, personal, interpersonal, intergroup and systemic, and could be used as a guide to diagnosing unsatisfied needs. For example, if an atmosphere of chronic lack of cooperation hampers a board's or committee's effectiveness, what might be the problem? If it is an apathetic lack of cooperation (see third column in Table 9.1) it could point to a lack of any sense of spiritual meaning – 'I don't really believe in what we are trying to do.' But if it is a more vigorous and obstructive lack of cooperation with a culture of blaming, then it could point to lack of personal or group acknowledgment – 'What's the point in trying if we won't get any credit?'

Rather than wait for destructive behavior to point the way, the following questions are designed to help you to explore your own needs and those of others so that problems can be anticipated before they become serious. At a

Table 9.1   Basic human needs and the potential effect of deprivation

Physical needs	Effects of deprivation	Destructive behaviors
Subsistence	Hunger, discomfort, and struggle for survival	Theft, fraud, unethical selling
Protection	Vulnerability. Must 'protect one's own'	Exploitation of natural resources Breaking environmental regulations Unacceptable working conditions Causing poverty and starvation
Financial security	Poverty	Addiction to making money
Physical health	Ill health, insomnia, aches and pains	Over-eating, alcoholism, workaholism
Mental needs	**Effects of deprivation**	**Destructive behaviors**
Understanding	Insecurity	Bad decisions because of lack of information
Stimulation	Boredom Restlessness	Good people leaving Sabotage
Challenge	Apathy	People allowed to get away with inexcusable behavior
Emotional needs	**Effects of deprivation**	**Destructive behaviors**
Affection	Fear, resentment, anger, anxiety	Betrayal, cheating, neediness, revenge
Trust	Guilt, hate, unhappiness	Betrayal, cheating, neediness, revenge, suspicion
Acknowledgment	Frustration, loneliness	Blaming and non-cooperation
Spiritual needs	**Effects of deprivation**	**Destructive behaviors**
Meaning	Meaninglessness	Apathy Justification to commit crimes
State of being	Cynicism, apathy, bitterness	Expressing discontent
Integrity	Guilt, self-doubt, not being trusted	Lying, deceit, manipulation, inconsistency, cheating
Social needs	**Effects of deprivation**	**Destructive behaviors**
Sense of belonging	Isolation, blame, lack of trust	Betrayal, insider dealing, leaking information
Support	'Go it alone' Victim mentality	Causing disruption and discontent Giving up
Participation	Frustration, loneliness	Sabotage
Loyalty	Hostility, apathy	Justification for undermining the group or people in the group
Need to give and receive love	**Effects of deprivation**	**Destructive behaviors**
Compassion	Anger, hurt	Tyranny, or over-control
Empathy	Lack of identification	Defensiveness
Intimacy	Self-centered, disrespect	Self-denigration, ruthless self-pursuit

Table 9.1 (continued)

Need to be creative	Effects of deprivation	Destructive behaviors
Freedom of expression	Loss of interest Frustration Depression	Wrecking initiatives Sabotage Silence
To be inventive	Depression	Creating low morale
To make a real difference	Hopelessness Apathy	Non-commitment
Identity needs	**Effects of deprivation**	**Destructive behaviors**
Self-worth	Low esteem Negative self-image	Driven behavior to get approval from others
Self-acceptance	Victim/tyrant	Bullying or colluding with bullying
Sense of self	Copy others	Insincere
Authenticity	Pretense	Generates distrust

personal level, the reader can take each question and rank themselves out of ten on the extent to which they feel their basic needs are fulfilled – from zero for utter deprivation to ten for total fulfillment. Then any score under five indicates an area where one should pay serious attention to an unsatisfied need. If you feel your needs are not being met, ask 'What can I do differently to help satisfy this particular need?' If you feel that it is unlikely that a need will ever be met or that your pattern of behavior is too deep-seated to change, at least you can let go of the expectation that ineffective attempts will fulfill your needs.

These questions could also be asked at other levels. At the interpersonal level they could indicate what a group of people needs in order to function well. They could also be asked by a group as well as an individual, for example, 'How comfortable are we with our physical surroundings?' Those making decisions impacting the intergroup and systemic levels could test whether any fundamental human needs were being undermined or satisfied as a consequence of their decisions. The tables could also be used in relation to setting ethical criteria for important decisions.

Bear in mind, however, that the answers to these questions are likely to be a mixture of what is real and what is perceived. If I am extremely wealthy, but I still feel financially insecure, there may be a real reason. On the other hand, I might be insecure because of childhood deprivation or pressures to do better than anyone else. It is important to differentiate between these real or perceived needs, rather than jumping to too hasty a diagnosis.

KEY QUESTIONS TO DETERMINE THE EXTENT
TO WHICH NEEDS ARE MET

Rate yourself out of ten the extent to which you feel satisfied. Ten is highly satisfied. Zero is not at all satisfied.

PHYSICAL

(a) How comfortable and at ease do you feel in your physical surroundings?
  • At work?
  • At home?
  • When traveling?
(b) How secure do you feel financially?
(c) Are you as healthy as you feel you need to be?

MENTAL

(a) Do you feel you have as much understanding of a situation as you need to make a decision?
(b) How fully do you think your ideas and feelings are really understood?
  • At work?
  • At home?
(c) To what extent are your needs for mental stimulation and challenge satisfied?

EMOTIONAL

(a) To what extent do you feel really wanted, appreciated and accepted?
(b) To what extent do you feel you can trust others?
  • At work?
  • At home?
  • With friends?

SPIRITUAL

(a) To what extent does life have a sense of purpose or meaning for you?
(b) How happy, content and at peace with yourself and your life are you?
(c) How certain are you that you are living your life honestly and with integrity?

SOCIAL

(a) To what extent are you supported in what you do?
(b) How satisfied are you that you are really making a contribution?
(c) Are you treated with respect and consideration?

(d) How loyal do you feel other people are to you?
(e) How loyal do you feel you are to:
  • Your work colleagues?
  • Your friends and family?

TO GIVE AND RECEIVE LOVE

(a) To what extent is your need to give genuine love satisfied?
(b) To what extent is your need to receive genuine love satisfied?

TO BE CREATIVE

(a) To what extent is your need to be creative satisfied?
(b) To what extent is your creative output appreciated and put to good use?

IDENTITY

(a) How well do you think you know yourself?
(b) Can you describe yourself as others would see you, in both positive and negative aspects?
(c) How true to yourself do you feel?

DIFFICULTIES ARISING FROM INEFFECTIVE MEANS OF SATISFACTION

Answers to the above questionnaire should give an indication of whether one's needs are being met or not, either in reality or as a result of a belief or perception. It is always worth getting a second opinion, preferably from an unbiased outsider, or psychotherapist, before taking such an analysis too far.

One of the problems about needs that are not consciously satisfied is that vicarious substitute behaviors can arise to fill the gap. Instead of creating contentment, these tend to create even greater feelings of deprivation. For example, someone who simply needs to be appreciated can become so demanding that others flee their presence. Unless the original need is satisfied in an appropriate way, it is likely that at a deep level the need will be diverted into ineffective means of satisfaction. Table 9.2 lists some of these 'negative satisfiers,' or ineffective ways we try to satisfy needs.

Understanding the nature of negative satisfiers and the underlying needs is a good basis for checking that individuals or groups are getting their real needs met. It can also be useful when thinking about the wider implications of corporate decisions.

Table 9.2   Negative satisfiers

Satisfier	Examples
Destroyers/violators	
These are paradoxical. Over time they annihilate the possibility of satisfaction and impair adequate satisfaction of other needs; e.g. a dictatorial chairman silencing his directors then asking why he has to make all the decisions	Imposed values Censorship Bureaucracy Authoritarianism Bullies and tyrants
Pseudo-satisfiers	
Elements 'generating a false sense of satisfaction of a given need' For example, the advertisement suggesting that some product will ensure a perfect happy family	Social persuasion Propaganda Financial or cultural promises Exploitation of natural resources Status symbols, cars etc. Advertising
Inhibiting satisfiers	
Those that generally oversatisfy a given need, at the cost of other needs. Often deep-rooted habits, customs and rituals For instance, you want to appear in a good light at a meeting so you constantly pass blame to others and this slows down the decision process	Smoking, drinking Workaholism Obsessions Paternalism/overprotection Unlimited permissiveness Obsessive economic competitiveness

If, by studying Table 9.1 or the questionnaire given in the previous section, the real need has been discovered and understood, this is in itself a big step forward because the worst excesses in Table 9.2 stem from needs that are unconscious and not understood. Such understanding also allows the time and energy that was wasted on negative satisfiers to be dedicated to finding more effective and satisfying solutions. For example, the bullying chairman in Table 9.2 might recognize from Table 9.1 that he is bullying his board because he does not feel accepted, that he was an outsider who did not feel that he was fully accepted so resorts to bullying to 'make his mark.' Although it is not easy to change one's behavior patterns, he can at least have a hold on the problem and can be motivated to let go of counter-productive behavior.

INAPPROPRIATE DEFENSES

Tables 9.1 and 9.2 illustrate how unfulfilled needs can lead to a wide variety of 'shadow-side' problems, but not all such reactions can be blamed upon one's own unfulfilled needs. Elements of the shadow side can creep in whenever there is stress or difficulty, because people need to defend themselves and do not always resort to appropriate defenses. Some misplaced defensive behavior can often unconsciously sabotage and undermine the ability of the individual and the group to do constructive work. This is a constant source of pain and irritation to me when I sit in on meetings, but it is understandable when one considers the complexities of human nature.

This study of patterns of defense is another useful way of understanding dysfunctional behavior. We are all interdependent on each other to make crucial decisions that affect our lives, and often those of many others. Because there is much at stake in terms of membership of any group, it is natural that defensive behavior is part of group dynamics.

Table 9.3 looks at some typical inappropriate defensive behaviors, in terms of questions to enable the reader to think about the kind of defense they might resort to. Perhaps you cannot identify directly with any of these, but you could ask others to give you constructive feedback if you are interested in finding out more. Try observing what your colleagues do when meetings are not going as well as they could – it is so much easier to analyze these problems in others.

Table 9.3　Ineffective defenses

DEFENSE	EXAMPLES	PRACTICAL SUGGESTIONS
Isolation When you are unhappy, do you isolate yourself from others? Do you keep things to yourself? Do you find yourself splitting off your thoughts from your feelings?	*Individual*: A chairman who is too busy to keep closely in touch with a company discovers that the CEO and directors are corrupt. *Group*: An executive team does not consult with its senior managers and is surprised when their policies are not implemented.	*For yourself*: If you find yourself withdrawing, try to let people know that you need time to yourself and that you will come back to them. If you do come back to an individual or a group of people after 'going private,' be sensitive to the negative impact it may have had. You may need to explain your thinking. *For another*: If you notice someone going private, point out to them that it has happened – it is very easy to withdraw without realizing the effect it has. Try to find out what the problem is. If this is not possible, either recommend that the person re-engages, or let them know they are welcome to talk about things when ready. This is particularly relevant to directors who do not contribute much in meetings.
Intellectualization Do you find yourself explaining a problem in purely analytical terms without considering your own or others' feelings or situations?	*Individual*: The analyst who gives an intellectual reason for withdrawing funds without considering potential human suffering he or she is likely to cause. *Group*: A board agrees a strategy that is highly ingenious without testing whether major customers will want their clever scheme.	*For yourself*: If you find yourself explaining things and notice your audience looking more and more unhappy, then logical argument may not be the issue; it could be practical, emotional or spiritual. Ask the person what the problem is, and try to understand. Then try to tune in and use more appropriate language; for example, acknowledge their feeling rather than explain it away. *For someone using intellectualization*: If someone is using intellectual arguments inappropriately let them know how you would prefer them to respond. You may have to repeat this request several times if they are in full swing.

Table 9.3 (continued)

DEFENSE	EXAMPLES	PRACTICAL SUGGESTIONS
Rationalization When you are challenged, do you feel you have to concoct seemingly plausible reasons to justify your practices and beliefs?	*Individual:* A non-executive director concocts a roundabout reason as to why he should discuss succession with a hostile chairman when he knows he is within his rights to do so. The chairman challenges his reasons and he sheepishly withdraws. *Group:* A board provides economic arguments to justify bypassing safety measures when they know it is wrong.	*For yourself:* When the person you are talking to looks impatient or loses interest, then try going straight to the point. *For someone who is using rationalization:* Ask him to say what he wants to say rather than the reasons for saying it.
Doubt and indecision Do you make efforts to avoid uncertain situations? Do people complain that you find it difficult to make up your mind or commit yourself to action? How often are you stuck because you doubt your own perceptions and views?	*Individual:* Share prices go down because an elderly chairman is taking too long to decide to retire. *Group:* A major acquisition is lost because the board keeps going round in circles and loses the opportunity to make an offer.	*For yourself:* Either ask at least four people what they would do in similar circumstances, and use their advice to work out a strategy for resolving the problem. Or, try to find out your real reasons behind the doubt or indecision and ask what is really important or what you really want. Ascertain your feelings. Ask your heart, 'Is it a yes or a no?' *If others suffer doubt and indecision:* Ask if they would like to talk it through. If they can't talk about it, and it matters that they make a decision, agree a deadline. Combat indecision with clarity – spell out exactly what will happen if they do not make a decision by then.

Table 9.3   (continued)

DEFENSE	EXAMPLES	PRACTICAL SUGGESTIONS
Denial Do you deny to yourself or others past facts or feelings that would be painful to remember? Do you instead focus on things that are benign or pleasant, pretending there is no pain, anticipation of pain, danger or conflict?	*Individual*: A chairman denied that he knew about irregular conduct among his directors when it was obvious from the figures. *Group*: An executive team denied that the sales director was incompetent and putting the company at risk 'because he was a nice guy.'	*For yourself*: If one person disagrees with you, it's chance; if two, it's interesting; if three people disagree with you it could be denial on your part. Observe whether you are putting your case over-vehemently, or if others are giving knowing looks. Rather than denying it, turn it round and ask yourself, 'If what I don't want to know is true, how could it be a useful learning opportunity for me to grow?' If appropriate, you could also check it out with others. Focus on facts. *If you think others are in denial*: Gently, patiently and in a non-judgmental manner, point out the facts. Repeat this until they get it. If, instead, you feel they are not in denial but consciously lying to hide something, prove your case and take appropriate punitive measures.
Projection How often do you know or do other people tell you, that you are blaming others for desires, feelings or faults that you have, or fear you have, yourself?	*Individual*: A CEO had a private lunch with a non-executive to complain about his chairman, whom, he said, was saying things about him behind his back. *Group*: A board blames market conditions for falling profits when they should have directed the CEO to find alternative ways of generating revenue.	*For yourself*: If you suspect you may be projecting a negative pattern onto another person, ask yourself, 'Who does this person remind me of? Has this happened before and is it part of a pattern? What hard evidence do I have that this person is really like that?' *If you think another person is projecting*: Check whether they are generalizing about types of people. If so, ask for evidence. If they are talking about a particular person, ask for hard facts, substantiated over time.

Table 9.3 (continued)

DEFENSE	EXAMPLES	PRACTICAL SUGGESTIONS
Regression How often do you avoid responsibility or unpleasant demands by evasive, wistful, dependent, ingratiating behavior that encourages others to indulge you instead of confronting the problem?	*Individual:* A CEO, who lied about the state of her company, appealed to the board because she was having problems at home. *Group:* A board resorted to humor and anecdotes to avoid being confronted with difficult decisions.	*For yourself:* If you feel needy, wanting attention, or resorting to manipulative means to get your way, stop. Do you feel real and authentic, or do you feel 'not truly yourself'? Address the need or anxiety behind the feeling. Learn how to ask for things directly or to state your thoughts directly. *If others are regressing:* If someone is acting childishly, do not respond to them in the same way, even though it is easy to do so. Bring them back to an adult level and find out what they are really trying to say, what they feel and what they want.
Displacement Do you feel dislike, anger or fear towards someone for no real reason?	*Individual:* Without thinking about it, a member of the nominations committee rejected every male candidate who was short. It transpired that when he was young, he had a sadistic schoolteacher who was short. *Group:* A board accused a newcomer that they did not know well of being untrustworthy and judgmental, when they were really describing the character of the chairman.	*For yourself:* Try the same methods as recommended for Projection. If you overreact to someone, question your motives. It may be Displacement. *For others:* Do not go along with judgments others make without checking that you feel that they are valid.

Table 9.3 (continued)

DEFENSE	EXAMPLES	PRACTICAL SUGGESTIONS
Reaction Do you react in a knee jerk manner to certain situations, for example being aggressive when you feel anxious? Or placating people to avoid confrontations?	*Individual:* Instead of discussing matters, a chairman would automatically say, 'If you don't like what I say, you are free to go. Plenty more where you came from.' *Group:* A board aggressively attacked an executive committee when they were anxious that it was not performing. The result was more defensive behavior and worse performance.	*For yourself:* As soon as you feel yourself going into a reactive fright or flight mode, pause. Pay attention to your breathing, and when your breathing is calm and even, reassure yourself that you are quite capable of dealing with the situation. If you have already reacted inappropriately to something and are aware of it, apologize and start again. *For someone who is reactive:* Realize that reaction usually comes from deep fear caused by a past trauma. Help the person feel safe before continuing the discussion.
Repression Do you sometimes suppress distressing feelings or avoid doing things you don't want to do? Do you avoid doing things you really want to do, like take time to think about things?	*Individual:* A chairman represses her frustration because a CEO is not performing well. She allows things to drag on for too long. *Group:* Non-executives are too polite to challenge either the CEO or the chairman, thus inhibiting the possibility of a good debate.	*Personal:* If you feel you are holding back on something, check whether this is either because of fear of repercussions or guilt. Ask yourself what you think and feel. Imagine you are the wisest person you know and do what you think that person would advise you to do. *For others:* If you feel people are repressing their genuine thoughts and feelings, ask them for their thoughts and feelings in private or in writing. Ask them, 'what conditions do you need in order to feel you can speak or act freely?'

PRACTICAL GUIDELINES FOR AVOIDING OR MITIGATING THE SHADOW SIDE OF CORPORATE GOVERNANCE

For those who do not have the time, need or inclination to work through each aspect of the shadow side in detail, the following general guidelines might be useful. To some extent, they echo suggestions already made earlier in the book.

1. Always respect your intuition. If you have a feeling that something is not quite right, you may be picking up something real. On the other hand, do your best to test it in case you are imagining things.
2. Always address any difficulty. If appropriate, talk it through with someone you trust. Then speak your truth from your heart.
3. Make sure you have a clear sense of right and wrong. If appropriate, make sure others are aware of what is, and what is not, acceptable behavior for you.
4. If someone, or a group of people, is behaving in a strange way, or you sense that they are, describe the behavior and ask them about it. If you still feel uneasy, continue to investigate until you are satisfied. If you are correct, deal with it as soon as possible. Do not procrastinate.
5. If dysfunctional behavior is occurring between you and another person, or in a group, and things are going badly, stop the interaction. Let people know that the interaction is not working for you. Let them know what would work for you. Ask them what would work for them. Work out a mutually beneficial way of sorting it out. Then proceed.
6. If the behavior of any person is either criminal or seriously undermines the safety or morale of the board, the executive or the organization, deal with it at once. Sometimes it is better for someone to leave promptly. You do not know how much more may be going on.
7. Welcome feedback about your own shadow side. If you recognize that it is true, then try to determine whether it is 'hard-wired' or can be remedied. Unless you decide that other things are more important at the time, apply the appropriate remedy. If what you are doing or not doing is damaging or hurtful to others, work out how you can mitigate its effects – admission and apology go a long way. Given that you do have weaknesses and blind spots, learn how to recover quickly and effectively. For example, in judo, when you are thrown, you learn to break your fall rather than hurting yourself.
8. Seriously ask yourself whether there are any basic needs that you have that require attention. Give yourself permission to have needs, and find healthy ways of meeting them. If necessary, find the courage to ask for them to be met.

9. Say no to anything that does not feel right to you. Also, never make a promise unless you commit to it in practice.
10. Always consider the human implications of any major governance decision you may make. If there are any negative effects on people, take these seriously into account and do your best to moderate the excesses of the downside. Be honest about them.

SUMMARY AND CONCLUSIONS

This chapter has tried to address some aspects of the shadow side of corporate governance and to give some practical suggestions for dealing with them. The subject is enormous, and no one chapter could do justice to it. This chapter gives just a brief outline of some of the most common problems found in corporate governance, such as the bully, people who suffer delusions of grandeur and the dangers of deceitful behavior. It was suggested that some of the major causes of the dark side of human nature stem from ineffective attempts to satisfy deep-seated human needs. It is hoped that the reader will go away with increased awareness and some practical suggestions about how to deal with the shadow side. For those interested in studying the shadow side more deeply, the books mentioned in Notes 4–7 at the end of this chapter make good reading.

Notes
1. Wilde, O., *The Ballad of Reading Gaol*, Dover Publications Paperback 1992.
2. Max-Neef, M.A. (1991) *Human Scale Development*, Apex Press.
3. Swanson, G.E. (1988) *Ego Defenses and the Legitimation of Behavior*, Cambridge University Press.
4. Masters, B. (1997) *The Evil That Men Do*, Black Swan.
5. Whiting, B.G. (1990) *Knights and Knaves of Corporate Boardrooms*, Bearly Ltd.
6. Wrangham, R. and Peterson, D. (1997) *Demonic Males*, Bloomsbury.
7. Storr, A. (1997) *Feet of Clay*, Harper Collins.

Part Four

The Human Face of Formal Roles

10 Corporate Functions and Formal Roles

▶ ▶ ▶   EXECUTIVE SUMMARY   ◀ ◀ ◀

This chapter describes the roles and functions of main or supervisory boards and executive boards or committees. Following discussion of the main or supervisory board it looks at the roles of chairmen and non-executive directors, then the executive board and the CEO, executive directors and the company secretary. Attention is also paid to interrelationships between individual roles, board and executive functions and investors. The chapter concludes with advice on ways to develop each role, with particular reference to the human side.

Part Two and the last three chapters have looked at some of the most obvious human aspects of corporate governance such as personal and interpersonal governance. However, the best personal and interpersonal governance is irrelevant if the groups concerned are not capable of fulfilling their formal and primary roles. This chapter looks at those roles and functions more directly concerned with the formal work of governing corporations. The functions and roles summarized in this chapter are gleaned from reports on governance and from actual observation of the work of successful chairmen, CEOs and their directors. The aim is to give an overview of the various governance tasks that people in their formal roles may have to perform.

People reading this chapter may recognize that they are good at certain subjects and not others. They may also feel that some of the criteria set are over-optimistic in terms of what can practically be achieved. The chapter will be of more help if the reader concentrates on those parts that are especially relevant or of particular interest.

The roles of the board and the executive committee are emphasized, while the role of the investor is merely touched on since it really belongs to another sphere. Before going into detail, Figure 10.1 summarizes what are considered to be the different functions of all three so as to put the rest of the chapter in context.

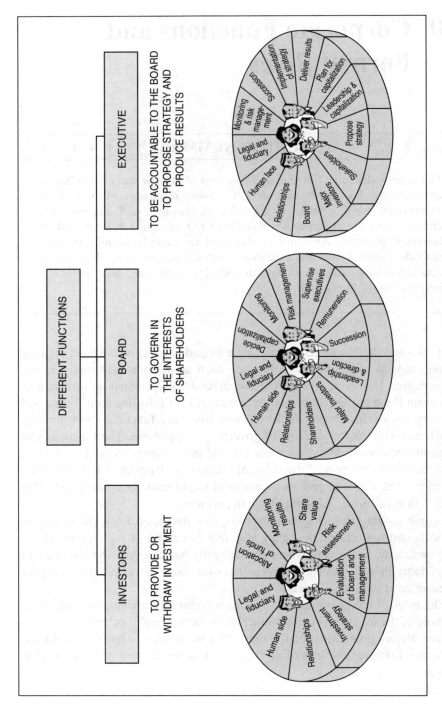

Fig. 10.1   The functions of investors, the board and executive committee

THE MAIN OR SUPERVISORY BOARD

This board has ultimate responsibility for making decisions about the fate of a corporation. At present, it acts in the interests of the shareholders and, in some cases, other stakeholders such as employees and the community in which the corporation operates. The way it operates depends on the structure and composition of the board, and especially on the nature and competence of the chairman, CEO and other board members. How effective it is depends on the competency of the board to carry out its functions. Its responsibility in terms of the human side will to a large extent be dependent on the commitment, values, principles and policies of the board and members' ability to put their values into practice.

This section outlines the most important of these functions as perceived by board members. They are explored in terms of the function of the board as a whole, the roles of the chairman, non-executive directors, the CEO and the company secretary. A brief description of the function of the executive board/committee and the roles of executive directors then follows.

What is good board performance?

This question has exercised the minds of many committee members, academics and consultants. Lists of competencies have been drawn up which aim to show the standards required.

However, one of the best definitions is a simple one.

> *An excellent board provides strategic oversight, evaluates corporate and top executive performance, represents and relates to shareholders, serves as a resource to top management, protects and enhances the company's assets and fulfils legal requirements. (1)*

It is generally understood that the board must first meet certain legal and fiduciary conditions in accordance with the legislation and regulations in place. However, extensive interviews with board members and observation of boards over many years suggest that this is only the starting point. In practice, there are multitudes of additional tasks – such as keeping the shareholders happy and taking responsibility for the board's own internal dynamics and succession planning – without regard for which, the board could not perform well.

THE MAIN FUNCTIONS OF THE BOARD

External relationships

Board members are often concerned that relationships with the outside world are maintained at an optimum level. The most important relationships are those with key investors, customers, the market as a whole, the government and, if relevant, with competitors.

The board is expected to ensure that legal and fiduciary requirements are met. It is also expected to be seen to understand and make key decisions concerning the strategic directions of the corporations it governs. Inspiring confidence, credibility and trust is a basic prerequisite of good governance. This includes demonstrating an awareness of environmental and ethical issues which are of public concern. It is highly probable that boards that seriously take the human implications into account will increasingly be viewed positively by investors.

Very often board members are appointed specifically for their contacts and networking skills. These also help the board to keep up to date with what is going on in the outside world.

In particular, key investors and shareholders must be kept happy. Because this is such a vital role of the board – especially for the chairman, CEO and finance director – it would be useful to address the question of what, other than good financial returns and share value, keeps shareholders and investors happy. Many shareholders are deeply concerned that they invest in companies that have a genuinely caring human face.

Keeping shareholders happy

According to most of the investors interviewed, the Organization for Economic Cooperation and Development principles give a good indication of what investors would like from boards. If a board comes up to these standards, investors will be happy. They are summarized below:

INVESTORS' PRINCIPLES OF BOARD CORPORATE GOVERNANCE.
OECD, 1999

1. **Fairness**
   That majority and minor shareholders receive their fair share of the firm's earnings and assets.

2. **Transparency**
    That there is timely disclosure of adequate information about a company's operating and financial performance and its corporate practice.
    - Financial reporting.
    - Ownership, structure and interests, including conflicts of interest.
    - Business operations.
    - Competitive situation. Potential areas of risk.

3. **Accountability**
    Ability of directors to provide independent oversight of management performances and hold management accountable to shareholders and other stakeholders.
    - Strong base of independent directors to look after the majority and minority interests of shareholders.
    - Clear roles for independent directors.
    - Composition and balance on the board.
    - Understanding conflicts of interest and ability to deal with these.

4. **Responsibility**
    That directors and managers meet legal standards and serve as responsible citizens.
    - Evidence of non-compliance with civil and criminal laws.
    - Evidence of corruption.
    - Track record of social responsibility. (2)

It would be interesting to see the extent to which boards believe that they truly meet all these principles, or indeed, check that they are meeting any of them. They look simple and common sense, but in practice it is often difficult to abide by them all. One director commented that 'All these rules and guidelines are all very well, but there is a big question about the how.' As suggested in Part Three, one of the main ways of doing this is through effective interpersonal governance, particularly efficient use of time, effort and talent.

Internal processes

The board, under the chairman, is responsible for carrying out its various duties either by one-to-one discussions with the chairman, or through various committees. The most common are the audit, remuneration, business development and nominations committees. The board also needs to be well represented at the AGM and is expected to have effective working relationships with key investors and shareholders.

The chairman is responsible for seeing that the board's activities are well managed, supported in this by the company secretary and the chairmen/women of the different committees.

The strategic roles of the board

One of the main roles of the board is to supervise the quality of strategic thinking of the executive committee, and make overriding decisions if it feels that the committee is not proposing the right strategy.

The board makes ultimate decisions about the fate of the company. This could mean rubber-stamping a decision, challenging ideas or pushing through alternatives. It is the executive's responsibility to come up with proposals for the board to agree on, or for the board to improve on. The board, therefore, has a key role to play in the leadership, direction and fate of that company.

It is also the responsibility of the board to ensure that the performance of the CEO and the executive meets the requirements of the shareholders. Good boards have a conceptual grasp of what is going on in the company as a whole, so that they can make sure that all the parts are in place. Mistakes are often prevented because some non-executives are vigilant and pick up danger signs and signals that need attention before they become problems. This means that they not only need to understand strategic content, but need to know enough about the companies they govern in order to make meaningful decisions about them.

Strategy is an umbrella term for many topics. I have discovered that directors mainly discuss strategy in terms of three main levels, which have been described as systemic, the structural/portfolio and the implementation levels:

- **Systemic-level strategy** is the thinking that relates to the board's awareness of what is happening in a national, international and global environment. This includes an understanding of social and political trends, awareness of potential market trends and the effects of changing technology. It also has to do with understanding international competitive forces. Boards also need to understand the present and potential competitive forces that might have a bearing on the company's future.

    Part of the work of the board is to establish and maintain good relationships with key parties such as governments, so that boards are aware of possible changes in the environment they are working in. Because executive directors are less likely to have as wide an external perspective as the non-executives have, this is one way in which some non-executive directors add value.

- **Structural and portfolio strategy** is what most boards think of as strategy. It has to do with the size of the company, the nature of the company, and what kind of business it thinks the company should be in. This includes the question of whether it should undergo restructuring and the level of capital expenditure. There are often discussions about the structure of the company and whether the company is able to support the growth aspirations of the board. Such discussions cover acquisitions, mergers and strategic alliances as well as the sale of companies or parts of companies.

- **Implementation strategy** is an area often neglected by the board. A vital role for the board is to ensure that it is satisfied that strategy is feasible. The board needs to ensure that an overall game plan for implementing policies and strategies is in place to ensure that management will actually deliver. Non-executives are in a good position to check whether all areas are covered or whether there are any gaps.

Quality and timelines of decision-making

No strategic thinking occurs in isolation. It occurs because there is a decision-making process which enables the board as a whole to make crucial decisions. The Hampel report argues that good strategic thinking requires 'high quality leadership.' As seen in the examples in Chapters 3 and 6, high quality leadership depends on the strength of the personalities. It also, however, depends on good decision-making. If this decision-making role is played well, boards are successful. If not, poor decisions are consistently made at the wrong times. As a result the board, and/or the company, is unlikely to last long. The quality of decision-making is key to good corporate governance. It is therefore one of the chairman's key roles to understand and master the art of decision-making. Guidelines for effective decision-making are given in Chapter 8.

In relation to decision-making, a major investor commented: 'The main role of the non-executive director is to ask the right questions. You need experience, intelligence, wisdom and shrewdness for that.'

THE ROLES OF THE CHAIRMAN

The overall role of the chairman is to manage the board and ensure that its policies are carried out. This includes working with the company secretary to address legal requirements, as well as understanding the financial implications of the company and keeping an eye on organizational targets and results. The chairman is there to ensure that good decisions are made with the aim of producing successful results.

In this sense, the chairman is a steward of board behavior and of company performance. To do this, as the chairman of one global company puts it, 'the perspective is all important. It is also important to grasp what really matters, and to get it across in a way that is understood. This is not always easy. It is a valuable discipline to have.'

Another chairman emphasizes the need to have an overview. 'The chairman's responsibility is the holistic future of the business. It must go in a sensible direction. Sometimes you need to have experience of competition and reacting expediently. At other times, you need to know when to hold things back.'

Part of the chairman's job is to have a general understanding of the way the company is run, how well it is doing and whether there are any gaps that need addressing. Many chairmen can sense when an issue needs to be discussed. They will often bring up a problem long before the CEO sees it coming. For example, one chairman predicted that sales revenues would fall within six months. 'I had a feeling about it, and when I checked it out, I was right.' By alerting the CEO and pointing out the implications, the CEO was able to take remedial action before sales figures started going seriously wrong. The buck ultimately stops with the chairman. 'If you get it wrong in certain areas, you get it wrong.'

The chairman also has responsibility for the external relationships referred to earlier – to ensure that the board acts within the interests of shareholders and, in some companies, of the stakeholders as well. Part of this means setting standards for the preparation of reports, writing the chairman's views, and the successful running of annual general meetings. It also involves establishing relationships with shareholders, major investors and other significant bodies such as the media and government.

A major part of the chairman's responsibility, however, is the internal needs of the board and its running. The chairman has to handle the human face of corporate governance more than anyone else does. Without adequate interpersonal skills, without being able to relate to, influence, coordinate and deal with a range of high-level people, it would be impossible to do this work. 'The board is a laboratory for human nature. It is all there.' A lot of the chairman's job is to understand and handle different characters, many of whom are powerful, strong willed, temperamental, and sometimes competing with each other. The guidelines on personal and interpersonal governance indicate that there is a great deal more to understand about individual board members and the board as a whole.

The chairman's key relationship is with the CEO.

Relationships with the CEO

Many chairmen believe that establishing good relationships with the CEO is crucial. 'It is up to the chairman and CEO to decide what is really important.'

For many chairmen and women, the relationship 'must be based on friendship. You need to get to know someone quickly, and establish a two-person team rather than a two-person division. If friendship is lacking, think seriously about what you are doing.' 'You need openness, trust and the right selection of issues to discuss.'

The chairman has to have a close working relationship with the CEO as he or she needs to have a view of the totality of what is going on in the organization, and to determine whether the CEO is meeting targets.

The relationship between the chairman and the CEO is a delicate one. 'No CEO does better than earning seven out of ten. What is required is recognition by the CEO that this is so. What is important is to deal with the missing three. If people clash or try to dominate, then you have trouble.'

It is true that some companies can be financially successful even though the relationship between the CEO and the chairman is unhappy. This is often at the cost of the mental and physical health of one or both. In some cases, the tensions are so bad that the health of spouses and children can suffer.

The disadvantage of poor relationships is that a CEO can get defensive and withhold information. Tensions and conflicts can disrupt board meetings and take up a disproportionate degree of board time and energy. Issues cannot be fully discussed. In addition, it often means that neither the chairman nor CEO is able to give of their best. Tensions and conflicts often cannot be avoided, particularly if a CEO is not performing well, or if the chairman thinks that an executive director should be dismissed. Both have to steer a careful path to avoid the situation deteriorating.

Relationships with executive directors

It is the CEO's job to manage his or her executive directors as part of an executive team. It is the chairman's role to ensure that the CEO is managing the team well. He or she sets the standards for deciding whether presentations and executive papers are relevant and well presented. 'What makes good non-executives are the executives. You need a team that understands the benefits of having to explain themselves. It's pretty rare.'

Relationships with non-executive directors

From the chairman's point of view, it is important to motivate non-executive directors and to find ways of eliciting their contributions. As one chairman put it, 'It is quite a feat to bring a non-executive back from the dead.' One of the problems has already been mentioned; that many non-executives do not have enough time. In fact about 40 per cent of chairmen feel that the current role of non-executives is a joke. 'As a chairman, I have to deal with the problem of enormous egos that need massaging.' In cases where the chairman has a good group of non-executives who are fully contributing, the board is not only successful, but the chairman creates a positive atmosphere. This means that a good chairman will attract some of the best non-executives.

Additional functions of the chairman within the board

The following indicates some of the key functions of the chairman in addition to those already discussed.

To set standards and make sure that policies and practices are in place

One of the chairman's tasks is to make sure that standards are at the level shareholders require, as well as being at the level needed for the board to function well.

This includes standards concerning information. Shareholders require adequate disclosure. The board needs the right quality, amount and timeliness of information from the executives for it to do its work. The chairman also sets standards of debate and behavior, in the boardroom and in terms of the running of board committees. He or she also needs to ensure that high-level policies are made, understood and maintained, not just about the business, but also in terms of social and environmental ethics. As one chairman suggested, 'Business is all about trade-offs. The art is to get the balance right.'

To ensure that directors make good decisions

Given that the board has to make the ultimate decision about what happens to a company, the ability of the chairman to ensure that good decisions are made is critical. In addition, the board needs to be able to deal with crises or to act urgently where it has to take advantage of major opportunities. The chairman has to lead the whole process when it comes to making major decisions. Some chairmen prefer decisions to be made outside board meetings and formally endorsed by the board. Others prefer open debate. This sometimes depends on the size of boards and the style and preferences of the chairman. The question of decision-making was covered in greater detail in Parts Two and Three.

Most board members will agree that 'usually, one person makes a decision. However, what is lost is the potential contribution a board as a whole might have made.' Because of shareholder pressure, there is a trend away from a dominant chairman to greater participation by directors. The chairman is now seen more as a facilitator and coordinator. (3) However, as one chairman commented, 'Boards can do some sensible things, but they are often too late.'

The whole issue of effective decision-making is one that concerns all directors, particularly as many feel that non-executive directors just do not have enough time to absorb material and to think deeply about critical issues. The speed and quality of decision-making is something that comes to many chairmen intuitively. Some chairmen are waking up to the possibility that there is more to understanding and mastering the decision process, including reconsidering just how much to expect from a non-executive director. Hopefully, the suggestions in Chapter 8 will go a long way to more effective use of time through conscious planning of the process and best use of individual board members.

To make sure that directors are continuously upgraded to the levels required by investors and which meet the current and future needs of the company

Chairmen are continuously asking themselves whether they have the right directors, and whether they are good enough. Some will ask whether there is any way in which they can improve the performance of the board. Reasons given for this include the argument that to keep ahead, it is necessary to improve the caliber of people. Also, that the future will demand different and better skills and expertise.

Successful chairmen make sure that the quality of their directors continuously improves and that those who leave are replaced by directors with the same or better qualities. They also have the wisdom and maturity to know when to leave and when to select and coach a successor. According to many board members, chairmen who know when to leave with good grace and preparation are thin on the ground.

To act decisively in times of crisis or when the company is at risk

I have observed that some chairmen and non-executives are at their best when dealing with crises. If there is trust in the chairman, and he or she acts promptly, disasters and unpleasant surprises are often avoided.

To act as representative and ambassador for the company

Depending on the relationship that the chairman has with the company and the shareholders, he or she often represents the organization. Chairmen and women attend meetings with key people, belong to trade and other associations, and talk to government officials and network. Many also enjoy the cut and thrust of doing deals, and will take part in high-level negotiations.

Many chairmen enjoy the challenge of the post. It is a mystery as to how all these tasks are fulfilled, given the limited time many chairmen and women have. The reality is that most chairmen and women are a complex combination of qualities and are better in some of these areas than others. There are many good chairmen/women, a few outstanding ones and many others who have the potential to be outstanding.

THE ROLES OF NON-EXECUTIVE DIRECTORS

The main role of all board directors, and especially non-executive directors, is to make decisions concerning the fate of the organization they are governing.

The main job of non-executive directors is to ensure that their individual and combined decision-making is of the quality and speed needed for effective governance. Not only do they need to be intelligent and have the relevant expertise, but also 'they need to have the courage to speak their minds, or they may as well be absent.'

In Chapter 7, A Model for Interpersonal Governance, it was suggested that directors could be effective both as participators and influencers. These are both art forms that can be considerably developed.

The kinds of decisions they may be involved in vary considerably. All directors have legal and fiduciary obligations. It is their responsibility to make sure that all activities are legal and meet regulatory requirements. In addition, the following responsibilities are important.

To provide objectivity

They provide external perspective about what is happening, both within the organization and outside. This is not possible for executives, who have their heads down and focus on running the organization. It is impossible for them to be totally objective.

An aspect of this is to oversee executive performance. Non-executives should be able to have a clear idea of whether a company is performing well or not. They should be able to read danger signs and signals well before a problem arises, and to indicate where they would like the CEO to pay attention. If this happened more often, more difficulties could be avoided.

To provide additional expertise and knowledge

In many cases, as a company changes or evolves, different levels of thinking, experience and expertise are required. For example, if a company is about to be floated, it is crucial to have a board member who knows and understands the financial markets.

To give clear directives in the case of crisis

Generally, the role of the non-executive is to challenge thinking, provide insights and perspective and suggest alternatives. The CEO can then decide whether or not to take this advice.

The CEO's responsibility is to meet targets. However, there are occasions when the non-executives believe that the company is not meeting targets or that poor performance is putting the company at risk. They may also fundamentally disagree with certain strategic proposals. In these cases, they have a duty to be prescriptive about what has to be done. It is also their ultimate duty to dismiss a CEO if they lose confidence in him or her. It is also their role to dismiss a chairman who loses their confidence, with permission from shareholders.

To recruit, select and dismiss the chairman, CEO and other board directors

The caliber of boards is one of the greatest concerns of investors. The difficulty is that it is not always easy for a board to know what criteria to set for the selection of new board members. The danger is that new members are selected either from the old boy network, or as replicas of the existing board. In many cases, the nominations committee does not have clear criteria or sufficient information about a candidate to make a real judgment.

Because this is such a difficult topic for some non-executives, hopefully the framework showing the transition from the past to the future in Figure 8.1 will be of use. Effective and timely succession planning is vital, but is done well in only a few companies.

THE ROLE OF THE EXECUTIVE BOARD/COMMITTEE

The role of the executive board/committee is to run the organization and to deliver agreed results. It is answerable to the board. It is also required to present strategic proposals, and proposals for capital expenditure. Once strategy has been agreed, it is responsible for setting targets and budgets, and for successfully implementing strategy. Its focus is on winning market share, producing goods and services, improving margins and reducing costs.

The executive committee, if it works well, operates more as a team. (4) Its intrinsic qualities, the team's 'inner brand,' affects the image it projects, both within the company, with the board and with shareholders and investors. This includes the team's values, beliefs, attitudes and behavior. These set the tone, morale and the culture of the organization. It also inspires or weakens confidence.

The executive group is only as strong as the ability of the team to work in an integrated way for the good of the organization. In this way, resources are maximized, the whole company works in harmony and executive directors are seen to be a powerful cohesive group sharing the same values, sense of purpose and ability to produce results.

THE ROLE OF THE CEO

The main role of the CEO is to lead, direct and manage his or her organization and to produce results. If expected results are delivered, everyone is happy. To achieve this, they need to cover the following.

To establish a constructive working relationship with the chairman

This entails clarity about the difference in both roles and expectations. It is also helpful if both the CEO and the chairman get to know each other's strengths and weaknesses well. They can complement each other and deal with any combined weaknesses and blind spots that might emerge.

A good working relationship includes a high level of trust and respect, and the ability to communicate with each other. Although some companies can be successful in spite of bad relationships, the cost both to effectiveness and to mental and physical health can be considerable. Time can be wasted on unnecessary conflict or avoidance of difficult issues. Tensions can result in the loss of good directors who find the environment distressing, or in a failure to attract good people. In many cases, tensions and conflict sabotage the ability of everyone concerned to focus on the tasks in hand.

To facilitate working with his or her directors so that they act in the interests of the whole organization

This group needs to agree policy and strategy, overall implementation planning and how to deal with issues such as IT, systems, structures, HR, and communications. It also needs to work together on problem solving, for example, around maximizing a major opportunity, resolving a crisis, cutting costs or improving company communications. The CEO must have the input, ideas and commitment of all of his or her directors to ensure delivery. If they operate as a group the chances of implementation are greater.

In addition to these

- To work with the executive directors on formulating strategic proposals to be endorsed by the board.
- To be involved in major deals and initiatives.
- Once strategy is agreed, to come up with a game plan to reassure non-executive directors that the strategic plan is achievable. This includes making sure that all the parts are in place.
- To provide leadership and direction for each of his or her executive directors.
- To be an ambassador with groups such as major investors, the media, government and other significant bodies. The CEO also represents the executive board to the non-executive directors. In addition, the CEO provides leadership, inspiration and direction to people in the company and to major customers and suppliers.
- To understand which aspects need attention, and to intervene where necessary to ensure good performance. The CEO needs to discriminate between allowing things to evolve and deciding when more radical measures are necessary.

THE ROLES OF EXECUTIVE DIRECTORS

Depending on whether an executive director is on the main board or not, there are two or three main roles. The first two are common to all executive directors.

To manage his or her own function or division

In the same way that the CEO manages the company as a whole, directors are responsible for managing their divisions and the people within them. They are responsible for submitting strategies and budgets, and competing for resources. They are also responsible for providing honest, accurate and timely information to the CEO. Ineffective communication of poor results has often cost both the director and, ultimately, the CEO their jobs. It is the role of the CEO to monitor the performance of each entity. However, the non-executives need to be kept informed about the organization to be able to detect danger signs and signals, or to ensure that major opportunities are not missed.

To work as part of a team for the overall success of the organization

The non-executive directors want to be sure that the company will be well managed. This means that executive directors need to take off their functional hats and think from the perspective of the company as a whole. This can be difficult, because there are sometimes conflicts of interests such as competing for resources. Executive directors need to switch mentally from operational thinking to conceptual thinking, and from running their own patches to working as a combined team if the company as a whole is to operate effectively.

The role of the executive director who is also on the main board

In addition to the above two roles, executives who are also main board members have the same legal and fiduciary duties as non-executives. They are also expected to make governance decisions that meet the interests of shareholders. Implicit in this is the assumption that directors are able to change gear from being an executive to a director, and be able to handle possible divided loyalties.

The advantage of executives being on the board is that fruitful discussions can take place between executives and non-executives. This increases the chances of better solutions and cooperation between the two. The problem is that often executives do not add value as directors, as they are afraid of showing lack of loyalty to the CEO.

THE ROLES OF THE COMPANY SECRETARY

The company secretary plays a vital part in board performance. He or she advises on legal and regulatory matters, and organizes the annual general meeting and dealings with shareholders. Most chairmen rely on the company secretary to check that the board is getting it right. Many company secretaries manage large legal departments. Some also play a large part in advising on negotiations and, in some cases, taking part in deals. He or she also writes the minutes of board meetings. In the way that they report meetings, they hold considerable indirect power.

Many company secretaries are also expected to keep abreast of thinking and practice about governance and to keep the chairman, CEO and other directors informed. Often, they also smooth out tensions between the chairman, CEO and other directors. As one company secretary put it, 'One of my main jobs is to act as an ego un-bruiser.' It is not just with the chairman and CEO, but also with other board members that the company secretary has to work. This means that they are in an excellent position to see how the board as a whole, and the executive board/committee, relate to each other.

Table 10.1 gives an idea of the interrelationships between the mainboard and the executive. If the reader is a member of a board or an executive committee, it should give an idea of where your company is in relation to others. If you are an investor or aspiring corporate governor, it may help you to discriminate between good, medium and poor performance.

Table 10.1 Where does your company stand?

	POOR	MEDIOCRE	GOOD	OUTSTANDING
MAIN BOARD	• Complacent or inadequate chairman running poor meetings, barely able to hold board together. • Poor balance between conceptual and operationally minded people. Little objectivity or external input. People in it for themselves. • Tension, conflict and lack of co-operation leading to breakdown of decisions. • Investors have low confidence because of poor results.	• Chairman going through motions. Allowing CEO to make decisions. • The CEO dominates decisions and misses crucial issues because no one challenges him. • Lack of trust and respect by executives for non-executives; non-executives complaining about executives. • Investors unsure. Results fluctuating.	• Chairman objective, keeping whole picture in mind. Able to balance various issues. Informs and briefs directors well. • High levels of debate. Board able to balance monitoring roles with making sure the quality of strategic thinking will produce strategy that keeps abreast of competition. • Mutual trust, respect, and honesty enabling constructive challenge and debate. • Good relationships with investors. Share prices steady.	• Chairman has high level of perspective, intellectual rigor, and emotional maturity. Understands and gets the best out of all his people. Inspirational. Ensures that timely and well-thought-out decisions are made. • Strategy is well presented by the executives, with key virtues emphasized. Execs and non-execs work together to produce better decisions as a whole than they would separately. • Genuine listening, learning and continuous improvement. High levels of ethical behavior. • Constructive, collaborative relationships with investors. Steady increase in shareholder value.
INTERFACE	• Incompetent CEO and chairman who cannot work constructively together. • Complaints by non-executives about the thinking and behavior of executives. Complaints that non-executives do not add value. • Low-level debate. Poor strategy. Self-interest more important than the company. • Unhappy relationships with shareholders. • Company at risk.	• Relationship between chairman and CEO sufficient to get the work done. • Non-executives do not have adequate information and time to make decisions. Executives go through the motions but do not act as directors. • Executives produce barely adequate strategy which is not well debated. Often views of CEO dominate. • Some relationships with investors good, others not. • Value/share prices fluctuate. Difficult to raise money.	• Relationship between chairman and CEO friendly, but not as rigorous as it could be. • All directors have their say. Their value is reasonably well understood – but the decision process is not well planned or thought through. • Executives produce reasonable strategy, well-debated, but not all directors contributing. • Executives and non-executives working together to ensure good relationships between analysts and investors. • Shares steady; company doing well.	• Constructive relationship between chairman and CEO. Agreement on key issues and how to handle them. • Excellent balance on the board. Directors all giving maximum value individually. Decisions made by the board as a whole greater than the sum of the parts. Collegiate atmosphere. • Brilliant strategy, recognized as such, Contingencies well aired. Well ahead of competition in thinking, timing and technology. • Excellent relations with major investors, shareholders and other stakeholders. • Share prices steadily rising.
EXECUTIVE	• Overconfident, dominant or weak CEO, undermining the ability of his or her people to deal with crucial issues. • No decision-making balance in the team. Too much focus on crisis management and operational virtues. • Fragmented team. Ego problems. Tension, conflict, lack of co-operation. Sabotage leading to poor or no decisions. • Low-level strategy and capacity to deliver. Poor results.	• CEO keeps things ticking over without making a significant contribution to the business. • Some balance between strategy and tactics, short- and medium-term thinking. Not well thought-through. • Team members put up with each other. Still operating as functional heads. • No clear understanding of strategy or key issues. Muddling though. • Results fluctuate. More a matter of chance, luck or historic circumstances.	• CEO has high levels of wisdom, knowledge and maturity. Good decision maker. Asks the right questions. Competent facilitator. • Reasonable decision-making balance and process. Focus on strategy and the big picture. • Directors committed to the good of the whole. Team working well but could do better. • Well-prepared strategy. In line with best practice. • Good results. Inspires confidence from investors, customers and other stakeholders.	• CEO as an inspirational leader and facilitator. Able to handle complexity, get the best from his or her people and continuously improve the quality and speed of decision-making. • Design process planned, taking into account individual strengths. • Directors able to work as one as well as deliver in terms of their functions. Operate as a high-level, powerful and cohesive team. • Inspirational strategy; differentiated; ahead of the competition; able to get value from non-execs to improve strategy. • Consistent improvement in share value.

BOARD AND EXECUTIVE DEVELOPMENT

As evidenced in Table 10.1, there is usually room for improvement. In Chapter 8, Guidelines for Interpersonal Governance, succession planning and selection of new members were related to competencies needed by the board to meet present and future needs. This is closely related to the structure and composition both of the main or supervisory board and of the executive board/committee.

Structure and composition

The purpose of structure is to provide a form in which governance can take place. The composition of a board is to a certain extent determined by its structure. Structure will also determine the dynamics of board behavior. It is a different experience to be part of a board consisting of 35 people who formally represent different bodies, and a board of six non-executives who are selected for their expertise. These differences were highlighted in Chapter 6. Given that both have the same basic decision-making role, the way that the chairman manages them is crucial.

What structure is in place depends upon national and international regulatory requirements and practices. Board structures vary across countries, and have different characteristics. In international companies there is often debate about the best country in which to locate the headquarters, because of possible tax advantages. In some boards, for example in South Africa, there is also discussion about which ethnic groups and nationalities should be represented. Another issue concerning representation is the presence of women on boards (see Summary and Conclusions).

Structures can reflect national characteristics and have implicit in them certain values. For example, there is a two-tier board in the UK where both non-executives and executives sit on the same board. In Germany, the USA and Japan, boards consist entirely of non-executives. Major stakeholders, such as banks, are often represented on these boards. Stakeholders are considered an important part of governance in Germany and Japan, where all parties have a vested interest in sustainability.

The structure also reflects social values, such as whether hierarchical structures are accepted or whether there is a leaning towards more democratic practices. In Chapter 7, A Model for Interpersonal Governance, it was suggested that leadership style can be more or less directive or facilitative, the second being more democratic.

The trend is to move away from autocratic leaders. For example, in the past in the UK, it was considered acceptable that power could rest in a person who was both chairman and CEO. It is now considered dangerous to accept that one person on their own can make balanced decisions in an increasingly complex world. Strong recommendations have been made that the structure is changed to separate the roles of chairman and CEO. The consensus is that the board as a group is better placed to make the right decisions. There has been a shift from autocracy to a more devolved form of governance that is becoming almost a formal requirement in countries such as the UK.

Structure makes a difference to the way boards and committees operate. However, composition is vital. If the right people are present at the right time, and employ effective decision-making processes, excellent judgments and results will follow. Whether these decisions take into account social aspects or not depends on whether those involved in making them care about the human implications.

Figure 10.2 shows unitary and supervisory board structures.

Developing the human face of corporate governance

In this chapter, the more formal roles and functions have been described. How, then, does the human side relate to these?

The answer is in a number of ways. For those interested only in the material gains or status of a company, it is still relevant. There is nothing wrong with the motivation to win or achieve the best. There is nothing wrong with competing to get to the top. All these are normal human motivations. Without these motivations, business would not be as successful as it is. However, this is not all there is affecting the quality of governance. As the chapters on personal and interpersonal governance indicate, poor working relationships between people do not make for good governance. The chapters on both personal governance and the shadow side suggest that money and status may not fulfil basic human needs, even for the person whose only dedication is wealth and making it to the top.

Even for the person unconcerned with the human side, the quality of working relationships makes a difference. Examples of the way negative relationships can affect performance were given in Chapters 3 and 6. Better personal governance has been shown to improve board and executive performance. More effective interpersonal governance not only helps to improve working relationships and the dynamics between people, but also improves the quality and speed of decision-making. Guidelines for personal and interpersonal governance are therefore relevant to everyone concerned with corporate governance, regardless of their values or beliefs.

For those who are also committed to improving the human face of corporate governance, the following questions might help their thinking when performing their various roles.

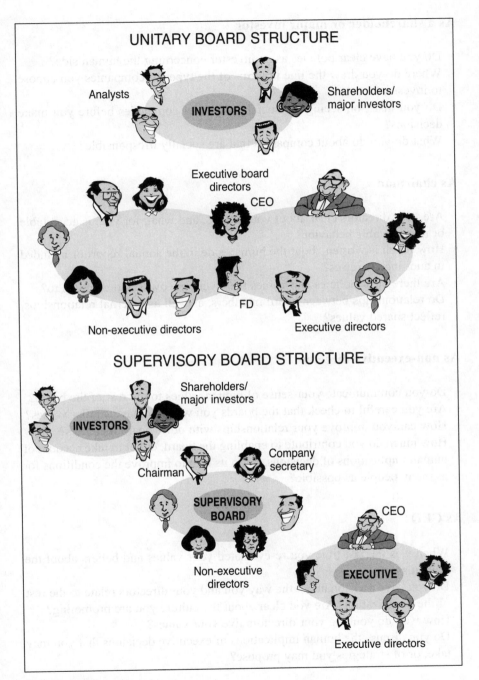

Fig. 10.2 Board structure

As a shareholder or major investor

- Do you have clear policies as an investor concerning the human side?
- Where do you draw the line in terms of the types of companies you choose to invest in?
- Do you insist on seeing the social policies of companies before you make decisions?
- What do you do about companies that are socially irresponsible?

As chairman

- Are your directors clear about your values, and what, for you, is acceptable or unacceptable behavior?
- How much is written about the human side in the annual report or included in auditing exercises?
- Are there clear policies and values that everyone owns and commits to?
- Do relationships between board members, as well as external relationships, reflect shared values?

As non-executive director

- Do you communicate your sense of human values to the rest of the board?
- Are you careful to check that the boards you work for have similar values?
- How can you improve your relationships with other board members?
- How much do you contribute to enabling the board, either to take account of human implications of decisions, or do its best to improve the conditions for as many people as possible?

As CEO

- When was the last time you re-evaluated your values and beliefs about the human side?
- Do you have a vision about the way you and your directors relate to the rest of the organization? Are you clear about the culture you are promoting?
- How well do you and your directors live your values?
- Do you discuss the human implications of executive decisions that you may take, or of strategies you may propose?

As executive director

- How favorably can you compare the way the company treats people with your values about the way they should be treated?
- If there is a shortfall, can you help to improve the situation? If so, how?
- Do you bring up issues concerning the human implications at meetings?
- How can you help people to work in a way that builds trust and respect, and is open and honest?

As company secretary

- Do you alert the board to state-of-the-art thinking about the human face of corporate governance?
- How often do you alert them to best practice in terms of relationships with shareholders?
- What would you do if the board you served was involved in practices that you strongly disapproved of?
- How often do you alert the chairman to the occurrence of dubious practices or potentially harmful practices?

It is hoped that these questions stimulate thinking in terms of how, as an individual, one could take the human side into account in terms of one's role, and how one operates as a human being. It is, however, the board or executive as a whole that makes the greatest difference.

SUMMARY AND CONCLUSIONS

This chapter outlines the main functions and roles of main or supervisory boards and executive boards/committees, and it suggests that personal and interpersonal governance are relevant to effective governance regardless of the values and beliefs of those involved. However, given the plea for making the human side an important part of governance, I would seriously question the right of those who govern to be solely interested in their own gain. They may be technically adequate, but may not really produce the kind of results that benefit mankind.

The following summarizes some of my desires concerning the behavior of corporate groups. Many corporate governors have similar wishes.

- To ensure that the quality of one's own – and one's own group's – working relationships genuinely express mutual respect, openness and honesty.
- To create a learning environment in which people feel they can give of their

best in practical, financial and material terms; intellectually, emotionally and spiritually, and support others to do the same.

- That people care deeply about the human implications of the decisions that are made.
- They have clear policies concerning what is, and what is not acceptable behavior.
- There is commitment to making a positive difference to the quality of the lives of all the people affected. Results reflect this at all levels, personal, interpersonal, intergroup and in the system at large.

If all these are practiced, what then would be the human face of corporate governance?

Notes

1. *The Family Business Adviser* (June, 1999), USA.
2. *Corporate Governance. Improving Competitiveness and Access to Capital in Global Markets* (1999). A Report to the Organization for Economic Cooperation and Development by the Business Sectors Advisory Group on Corporate Governance.
3. Garrat, B. (1997) *The Fish Rots From the Head*, Harper Collins Business.
4. Katzenbach, J.R. and Smith, D.K. (1993) *The Wisdom of Teams*, Harvard Business School Press.

Summary and Conclusions

SUMMARY

This book is intended to stimulate thinking and offer practical guidelines for those who want to improve the human aspect of corporate governance.

Part One reflected on what the human face means. Chapter 1 explored the social and technical climate in which chairmen/women, CEOs and directors may have to govern in the not too distant future. Chapter 2 argued a case for taking human sides of corporate governance more seriously. To do this would not only improve the quality of corporate governance in general, but could well be the next stage in its evolution. Business has mastered the art of making money, technology and communications. The suggestion was that if knowledge and expertise about the human side were combined with financial expertise, the world would be a better place. This includes improving the quality of life for those who govern, particularly in the areas of emotional and spiritual fulfillment. Four levels of governance were discussed – personal, interpersonal, intergroup and systemic. All these could benefit from improvement, but the book focused primarily on two levels, personal and interpersonal, as the foundation for the other two.

Part Two was about personal governance. Chapter 3 illustrated how different leadership styles can impact the nature of corporate governance. Chapter 4 offered a simple working model of self-governance, followed by guidelines for personal governance in Chapter 5. In both Parts One and Two it was suggested that effective governance is a combination of physical (including financial), mental, emotional and spiritual intelligences. These enable decisions to be made and results generated, but alone they are not enough. It is important also to consciously re-evaluate values and beliefs about the human side at both an individual and an interpersonal level.

The next section of the book, Part Three, explored interpersonal governance in a similar way. Chapter 6 described how different groups operate, while Chapter 7 offered a model that could be seen as both linear and dynamic. Chapter 8, Practical Guidelines, gave ideas for improving group decision-making and dynamics, while Chapter 9 was devoted to the shadow side of corporate governance.

Finally, Part Four was a short section linking the book to the more formal roles and functions of main and supervisory boards and executive committees.

It is important to note again that the stories described in Chapters 3 and 6 are purely fictional. Any resemblance to any person or group of people is therefore coincidental.

CONCLUSIONS

By now it should be clear that corporate governance is nothing if not human.

In one sense, human nature does not seem to change at all. We go about our daily lives, loving and supporting, or hating and destroying, each other. We are highly creative beings capable of running large as well as small empires. What we actually create depends on our values, beliefs, our state of being and the extent to which we unconsciously act out our destructive shadow sides. It also depends on how we maintain or change the systems we have inherited. In another sense, human beings have indeed evolved, but often in separate areas. Knowledge and expertise about the human side has now reached a point where people involved in corporate governance can take advantage of it. It is now quite possible for individuals and groups of people to make a significant difference if that is what they want. For those who argue that corporate governance is only about generating financial wealth and making shareholders happy, I would suggest that they lag far behind current thinking.

The combination of financial, technical and commercial expertise, compassion and greater awareness and mastery of self and interpersonal governance can be extremely powerful. I have not only experienced it myself, I have witnessed it on many occasions. I know it is possible both to generate wealth and at the same time benefit many people in terms of the non-financial aspects of their lives. It has to do with the basic humanity of the group that is making key decisions. Because corporate governors have power over the fate of a corporation, it also applies to the humane treatment of staff when corporations are shut down.

People are often so busy carrying out their tasks that they sometimes forget to think about the human implications of decisions they make or the communities they impact. In the seduction of figures, it is not always easy to see statistics in terms of actual human beings with faces and personalities. If one was to stand back and put oneself in the place of those most deeply affected, the decision might be very different.

The importance of self-development

First and foremost, the human side begins with how each person perceives and values themselves and others. In my experience, younger ambitious CEOs and chairmen are beginning to see the merits of self-governance and are seriously spending time on developing self-awareness and questioning their values.

The simplest and most difficult gift that anyone could give is to really find and speak the truth, particularly to oneself. It always upsets me when I am working with a well-known powerful and extremely wealthy person who says, 'If I learn too much about myself, I am afraid that there will be things I don't like about me.' The answer is yes, there probably will be. There will also certainly be positive aspects that can be better developed. The sooner we give up the expectation that we should be perfect, the easier it is to deal with one's own and other people's destructive and negative sides.

Self-development is not a self-indulgent luxury or something needed for people in emotional distress. Without it, corporate governors tend to work semi-unconsciously and in limited ways. As most people know, if one does not pay attention to one's own self-development, life sometimes comes along and forces one to take stock. How evolved a person is will to some extent determine how well they cope.

When other people in a group also commit to self-development, they become more effective and more aware of how best to use their strengths and handle their weaknesses. It often has a positive effect on the group as a whole, particularly in terms of the quality of working relationships. In addition to technical and subject matter expertise, what is also needed is wisdom and compassion. This comes from the ability to be true to oneself and to be more effective by integrating the physical, mental, emotional and spiritual aspects of oneself.

The importance of improved interpersonal governance

More leaders are also consciously spending time on creating better working relationships with their groups. In the past the practice was to bring in qualified facilitators. While this is often still necessary, it is now time for more leaders to acquire the skills and knowledge for themselves.

The challenge is to significantly improve the speed and quality of decision-making by developing the skills and expertise to do this. As time becomes increasingly precious, corporate governors cannot afford to waste the amount of time that they currently do. Those leaders who combine wisdom, intelligence, compassion and effective leadership processes are the ones that enable others to produce outstanding results. This also applies to board and committee members who contribute either by participating, influencing or both.

It is my belief that with time, training and effort, most governance groups could considerably improve their performance. Because I have seen it work, I know it is possible to create an environment in which people work together in profoundly enjoyable and satisfying ways. I once asked a religious leader what changes he would like to see in the communities he served, and he replied 'smiling faces.' He was concerned that people were not as happy as they could be. 'We seem to be suffering from spiritual dereliction.'

This unhappiness is evident in many boardrooms and executive groups. Some people may be hard-wired to be miserable, but this is not so for the majority. If this is an attainable state of being, why should meetings be boring and miserable?

Intergroup governance – the next stage

In the main body of the book, I have not talked a great deal about intergroup governance. There are two reasons for this.

The first is that individuals and groups need to get their acts together first. Intergroup governance is, after all, interpersonal governance between people of different groups. However, having said that, what is helpful is for each person to seriously consider two things. The first is whether you are harnessing your strengths in the right way with the right groups to get maximum effect. Some people are better at some levels than others. In spite of complaining of lack of time, few of us stand back and consider whether the way we rush from group to group might be fragmenting the best that we can offer. Often the most difficult skill is to feel OK about saying 'no.'

The second aspect refers to the people with whom you associate. On the one hand, one can only deal with the people in front of you. On the other hand, if you are working with people you fundamentally distrust, do not respect, and whose company you do not enjoy, is it an environment that is right for you? It may be your mission to help people transform, but most of us give of our best when we work with people who really support us. How many of us actively decide to find the right people?

If it is not possible to choose the people you are working with, perhaps networking is the answer. Networking has been a popular practice for some time. Some people are extremely good at it. For those who are committed to improving the system in terms of rules and regulations and monitoring, then the art of networking and working with influential groups needs to be taken to another stage of development. We need to become more effective in the ways in which both global and social policies and practices ensure greater humanization of corporate governance. Rapid technological change is likely to make this both easier and more of a challenge, given the scope for greater abuse as well as potential benefits. It may therefore be necessary to anticipate the human implications well in advance and for influential groups of people to move faster. People well know that the earlier an issue is dealt with, the less time there is to establish deeply entrenched behavior.

Finally, I shall complete the book in the same way that I began it, on a personal note. In terms of the human face of corporate governance, it is important to say that I do not feel that I have the answer. Hopefully I have asked some worthwhile questions and given the reader some ideas and practical suggestions for finding their own solutions.

The role of women in corporate governance

One area which requires an answer and which is close to my heart is the role of women in corporate governance.

In spite of the fact that there are more women on executive committees and boards and that many women have proved their capacity to govern, there are still fewer women in positions of power than might be expected. There are many reasons for this. A key reason is that, apart from inertia, apathy and fear, there do not appear to be clear reasons why boards would benefit from more women members.

What I do know is that when there are women on boards, the atmosphere usually becomes more cooperative and co-creative. Groups of men on their own tend to play pecking order power games – as a female CEO of a large research organization put it, 'too much testosterone.'

It has been proved that women are capable of fulfilling roles as business leaders running successful corporations and being brilliant investors. Although it is dangerous to generalize, there may also be certain contributions that women bring that are lacking in many boardrooms and committees. This is not to say that no men have similar qualities: some do, but not enough to provide the right kind of balances for effective decision-making.

Women tend to bring different perspectives and ways of seeing things. They tend to be more aware of emotions and find it easier to express and deal with feelings than many men do. Many of the women leaders I have observed have a shrewd understanding of what is going on with people in their group. They are less likely to tolerate dysfunctional behavior.

Many women also have a lot to contribute in terms of improving the human face at the systemic level. They often have a closer feel for communities affected by corporate decisions. Many are more concerned for the implications of decisions on families and children than those men who find it easier to segment off work from home.

Equally important, the market must serve the total population. If women are not represented in corporate governance, boards and committees are not well represented. They are missing out on 50 per cent of the adult consuming population.

These are starting points for why there should be more women in corporate governance. What I would like to see is this subject taken more seriously. It would be good to have a series of case studies indicating the benefits of more women in corporate governance. Even more important would be the benefit of having honest and profound dialogue between men and women to create real understanding and co-create effective ways forward. There are many issues that need to be resolved, including removing prejudices and negative expectations on both sides.

There is also the question of how to manage sexual energy and the increased atmosphere of sexuality that can be sparked off. Powerful, rich, positive and successful people can be extremely attractive, both men and women. A number of men have told me that for this reason, they are scared to have women involved in corporate governance, particularly powerful ones.

One of the main issues to come up in discussions is how to deal with potential power conflicts inherent in male/female dynamics. However, power games are not just about the battles of the sexes.

The issue of competition

Power games are the fabric of our system of corporate governance. They are particularly endemic in a society that encourages competition as a major value. There is much merit to competition: it enables the best or the most astute to win; it encourages improved performance and standards; it is an effective way of choosing between alternatives; and it encourages people to be on their mettle.

However, there can be a human price to pay, not only for the individuals concerned, but also for people within the system as a whole. Because of this, I would like to see more dialogue by business leaders on this subject and greater direction given as to how competition and the human side can be better bedfellows. Shareholders and major investors as well as boards and executives should also have their say.

Finally, I hope that this book has indicated that to think about the human face as part of corporate governance raises profound questions about the true nature and scope of corporate governance.

There are numerous examples all over the world of people involved in corporate governance and making a difference to the human side. For example, BP's policies on pollution had a major effect on the oil industry. Many corporations are helping deprived communities in their attempts to regenerate the economy. The Nagarjuna Group in India not only has enthusiastic employees, but has also created a beautiful natural environment and significantly helped to improve the lives of the local farmers and small industries. People like George Soros, Bob Monks and Nel Minnow amongst others are creating ripples in the investment world that will significantly change the human face of corporate governance.

Examples like these prove what can be done and show what is actually being done. There is, however, potential for much more to be achieved. Hopefully this book will have made a small contribution.

Index